THE FUTURE IS PRESENT

Leonardo

Seán Cubitt, Editor-in-Chief

Ecstatic Worlds: Media, Utopias, Ecologies, Janine Marchessault, 2017
Interface as Utopia: The Media Art and Activism of Fred Forest, Michael F. Leruth, 2017
Making Sense: Cognition, Computing, Art, and Embodiment, Simon Penny, 2017
Weather as Medium: Toward a Meteorological Art, Janine Randerson, 2018
Laboratory Lifestyles: The Construction of Scientific Fictions, edited by Sandra Kaji-O'Grady, Chris L. Smith, and Russell Hughes, 2018
Invisible Colors: The Arts of the Atomic Age, Gabrielle Decamous, 2018
Virtual Menageries: Animals as Mediators in Network Cultures, Jody Berland, 2019
From Fingers to Digits: An Artificial Aesthetic, Ernest Edmonds and Margaret A. Boden, 2019
MATERIAL WITNESS: Media, Forensics, Evidence, Susan Schuppli, 2020
Tactics of Interfacing: Encoding Affect in Art and Technology, Ksenia Fedorova, 2020
Giving Bodies Back to Data: Image-Makers, Bricolage, and Reinvention in Magnetic Resonance Technology, Silvia Casini, 2021
A Biography of the Pixel, Alvy Ray Smith, 2021
Living Books: Experiments in the Posthumanities, Janneke Adema, 2021
Art in the Age of Learning Machines, Sofian Audry, 2021
Design in Motion: Film Experiments at the Bauhaus, Laura Frahm, 2022
Northern Sparks: Innovation, Technology and the Arts in Canada from Expo '67 to the Internet Age, Michael Century, 2022
The Artwork as a Living System: Christa Sommerer and Laurent Mignonneau 1991–2022, edited by Karin Ohlenschläger, Peter Weibel, and Alfred Weidinger, 2023
Computational Formalism: Art History and Machine Learning, Amanda Wasielewski, 2023
Picture Research: The Work of Intermediation from Pre-Photography to Post-Digitization, Nina Lager Vestberg, 2023
Art + DIY Electronics, Garnet Hertz, 2023
Voidopolis, Kat Mustatea, 2023
After Eating: Metabolizing the Arts, Lindsay Kelly, 2023
Tactical Publishing: Using Senses, Software, and Archives in the 21st Century, Alessandro Ludovico, 2023
The Future is Present: Art, Technology, and the Work of Mobile Image, Philip Glahn and Cary Levine, 2024

See http://mitpress.mit.edu for a complete list of titles in this series.

THE FUTURE IS PRESENT

ART, TECHNOLOGY, AND THE WORK OF MOBILE IMAGE

PHILIP GLAHN AND CARY LEVINE

THE MIT PRESS
CAMBRIDGE, MASSACHUSETTS
LONDON, ENGLAND

The open access edition of this book was made possible by generous funding and support from MIT Press Direct to Open.

The MIT Press would like to thank the anonymous peer reviewers who provided comments on drafts of this book. The generous work of academic experts is essential for establishing the authority and quality of our publications. We acknowledge with gratitude the contributions of these otherwise uncredited readers.

This book was set in Arnhem Pro and Frank New by Westchester Publishing Services.
Printed and bound in the United States of America.

Library of Congress Cataloging-in-Publication Data

Names: Glahn, Philip, author. | Levine, Cary S., author.
Title: The future is present : art, technology, and the work of Mobile Image / Philip Glahn and Cary Levine.
Description: Cambridge, Massachusetts : The MIT Press, [2023] | Series: Leonardo | Includes bibliographical references and index.
Identifiers: LCCN 2023030210 (print) | LCCN 2023030211 (ebook) | ISBN 9780262548076 (paperback) | ISBN 9780262378734 (epub) | ISBN 9780262378727 (pdf)
Subjects: LCSH: Mobile Image (Artists' collective)—History. | Technology and the arts—History—20th century. | Art and society—History—20th century.
Classification: LCC NX512.M58 G59 2023 (print) | LCC NX512.M58 (ebook) | DDC 709.2/273—dc23/eng/20231206
LC record available at https://lccn.loc.gov/2023030210
LC ebook record available at https://lccn.loc.gov/2023030211

10 9 8 7 6 5 4 3 2 1

CONTENTS

SERIES FOREWORD

Leonardo/The International Society for the Arts, Sciences and Technology fosters transformation at the nexus of art, science, and technology because complex problems require creative solutions. The Leonardo Book Series shares these aims of artistic and scientific experimentation, and publishes books to define problems and discover solutions, to critique old knowledge and create the new.

In the early twentieth century, the arts and sciences seemed to interact instinctively. Modern art and modern poetry were automatically associated with relativity and quantum physics, as if the two were expressions of a single Zeitgeist. At the end of the World War II, once again it seemed perfectly clear that avant-garde artists, architects, and social planners would join cyberneticists and information theorists to address the problems of the new world order and to create new ways of depicting and understanding its complexity through shared experiences of elegance and experiment. Throughout the twentieth century, the modern constantly mixed art and science.

In the twenty-first century, though, we are no longer modern but contemporary, and now the wedge between art and science that C. P. Snow saw emerging in the 1950s has turned into a culture war. Governments prefer science to arts education yet stand accused of ignoring or manipulating science. The arts struggle to justify themselves in terms of economic or communicative efficiency that devalues their highest aspirations. And yet never before have artists, scientists, and technologists worked together so closely to create individual and collective works of cultural power and intellectual grace. Leonardo looks beyond predicting dangers and challenges, beyond even planning for the unpredictable. The series publishes books that are both timely and of enduring value—books that address the perils of our time while also exploring new forms of beauty and understanding.

Seán Cubitt, Editor-in-Chief, Leonardo Book Series
Roger F. Malina, Executive Editor, Leonardo

ACKNOWLEDGMENTS

First and foremost, we would like to thank Kit Galloway and Sherrie Rabinowitz, without whom this book would not exist. Kit has been enormously generous with his time and ideas and in giving us access to his magnificent archive. We only wish Sherrie was here to see the book. We would also like to thank Ben Caldwell, Ana Coria, Ulysses Jenkins, Daniel Martinez, Hyesook Park, Jose Luis Sedano, and Gene Youngblood for the hours they spent with us discussing the work of Mobile Image and the broader Los Angeles community. The following people were vital to our research and publishing: Megha Bhardwaj Azad, Grant Bollmer, Rebecca Brown, Seán Cubit, Jay Curley, Emily Elliott, Ellery Foutch, Gabriela Bueno Gibbs, Mark Hineline, Luane Hutchinson, Aric Mayer, Michael Jay McClure, Emily Pugh, Jordana Moore Saggese, Sarika Sanyal, Yiğit Soncul, James Toftness, Andrea Truitt, and Hélène Valance. This project also received crucial institutional support from the Getty Research Institute, the Museum of Contemporary Art, Los Angeles, the Tyler School of Art + Architecture's Deans Grant program, the University of North Carolina, Chapel Hill's Carolina Seminars Program, and UNC's Department of Art & Art History, Institute of Arts and Humanities Faculty Fellowship Program, and Publication Support Grant. We would also like to thank our students and colleagues at Tyler and UNC, along with the following people for their indispensable help along the way: Alexander Alberro, Nora Alter, Michael Connor, Aria Dean, Erin Dickey, Jonathan Furmanski, Matthew Jesse Jackson, Richard Langston, Alyse Levine, Nell McClister, Eric Muller, Kristen Neville Taylor, Mark Simpson-Vos, and Ben Tiven.

INTRODUCTION: TECHNICS OF AESTHETICS

It takes more than electronics to make a network function.
—Mobile Image[1]

This is a book about art and technology. It is a critical history of the prescient artist group Mobile Image, whose work with communications, networking, and information systems in the 1970s and 1980s offered profound lessons that have become even more pertinent today. Founded by Kit Galloway and Sherrie Rabinowitz in 1977 and employing all kinds of then-futuristic devices, from satellites to databases to electronic message platforms, Mobile Image appropriated emerging technologies in ways that creatively reacquainted individuals and groups with the means of producing and disseminating knowledge and experience, revealing latent potentials for emancipatory use. Based in Los Angeles, the group operated amid economic precarity, urban reorganization, a techno-boom, media consolidation, and ascendant neoliberal politics. In the long wake of the feminist and civil rights movements, Mobile Image epitomized the ever-intensifying conflict over the representation and mediation of bodies, identities, and communities.

This is also a book about art *as* technology. It proposes that the critical and creative act of extending the self into the world—and vice versa—has the potential to examine, reimagine, and functionally reorganize existing relations between people, objects, and environments. Thus conceived, art becomes a process that both connects and distinguishes subject and object, a process whose politics comprises the struggle over degrees to which it liberates and exploits. Consciously engaging in that struggle, in a participatory practice that revealed the potential for radically new forms of exchange among diverse publics, Mobile Image extended the project of the historical avant-gardes to envision and activate the transformation of society collectively through the tools at its disposal, not just those traditionally reserved for artistic practice but potentially all instruments of material and immaterial production and exchange.

Focused on one of the most visionary and under-studied artist collectives in postwar history, this book poses a broader challenge to entrenched histories of contemporary American art, along with the politics involved in the construction of those histories. The disregard for artists such as Mobile Image is symptomatic of the depoliticization of art history that emerged in the aftermath of World War II, as matters such as medium specificity, individual creation, and autonomy increasingly took precedence over questions of function, utility, audience, and experience. Mobile Image charted a different path, in which technical innovation and political innovation were one and the same; indeed, the latter was the criterion by which the former was to be judged. And yet, the group's recuperation of avant-garde tactics required a new strategy, responsive to the cultural politics of the moment—not only the immediate needs of the individuals and communities with whom they worked but also the general rejection of teleological outlooks and grand narratives in the 1970s and 1980s. The question of how to resist and reimagine the current order without replicating its oppressive logic was paramount.

Mobile Image executed three key works between the mid-1970s and mid-1980s: *Satellite Arts (The Image as Place)* (1977), *Hole in Space (A Public Communication Sculpture)* (1980), and *Electronic Café* (1984). Years in the making, and building on one another in a process of continuous research, each took an incisive approach to innovation while perceptively reflecting the material and immaterial complexities of its moment. These works were designed to creatively redeploy existing but futuristic-seeming devices as well as their techno-logics, to demystify and make available the latent powers of such tools and their users. *Satellite Arts* enabled people to interact physically across a three-thousand-mile divide; the work explored and expanded the always-already mediated corporeality of bodies in televisual space. *Hole in Space* linked public locales and everyday passersby in Los Angeles and New York City, an experiment in transcending the economic and psychological boundaries of urban space. *Electronic Café* was a seven-week-long network of neighborhoods and communities spread across LA; it was revolutionary in its innovative production of new social imaginaries, consciousnesses, and formations. Each chapter of this book focuses on one of these works, its context, and its relationship (and contribution) to interconnected discourses and practices of art, technology, and politics. Together, these works form an argument for the historical importance of Mobile Image and for a critical process that is at once analytic and transformative, that challenges core conventions of the public sphere, democracy, communication, and political participation, and that seeks to reframe myths of progress and the relationalities of power, representation, and identity—and thus change them.

Throughout all three projects, Mobile Image was committed to providing the agency, technology, and know-how to transform the material and ideological conventions and constrictions of the past and present—bodies, spaces, images, actions, knowledges, and experiences, the tools of individual and social selfhood. The group tapped into what Walter Benjamin had once enthusiastically termed "the desire of the present-day masses to 'get closer' to things spatially and humanly, and their equally passionate concern for overcoming each thing's uniqueness."[2] Within the context of a post-1960s United States and its increasingly neoliberal economic and ideological orders, Mobile Image employed state-of-the-art TV-network technologies, audio and video equipment, searchable text and pictorial databases, image exchange and audio-conferencing equipment, and digital writing and drawing tablets, aiming first and foremost to forge connections among and toward people, things, ideas, and the devices themselves in order, in the Benjaminian sense, to get closer, to know, and to take up anew, differently, imaginatively.

Understanding subjectivity as a set of fragmented, intertwined, and often conflicting relations between public and private, margin and center, Mobile Image sought the means to repossess and redistribute communication technologies productively in ways that foregrounded such relations, along with the particular circumstances and histories that determine them. Their work modeled a new *technics of aesthetics*, an active, historically specific, and dialectical engagement with such tools and their roles in the construction and distribution of perception, in the economy of images and imaginaries. The goal of such an approach is to see and allow others to see where and how one is positioned within a system of relations—where and how one can be, ought to be, wants to be.

Inspired by the work of Mobile Image, their friend and collaborator Gene Youngblood wrote in 1986: "The concept of an avant-garde, disavowed by postmodern theory, is actually more relevant today than ever before, but it has nothing to do with aesthetics. Only social situations, not artworks, qualify as avant-garde. We need access to alternative experience, not merely new ideas, for we know more about our being than we have being for what we know."[3] This book draws on particular aspects and examples of the so-called original avant-gardes in part because of the particular moment to which the works in question belong: as a general wariness of the marriage of art and revolution increasingly took root, certain artists, critics, and historians sought connections between movements and aspirations of artistic practices committed to the radical transformation of everyday experience, critically investigating the term "avant-garde," its legacy, and legitimacy.[4] The work of Mobile Image foregrounded communications technologies as aesthetic tools that would reorganize the social construction of realities, decentralizing and pluralizing a media infrastructure that had inspired many works of art but remained largely unaffected by an art world that largely prioritized the work in its closed and thus exchangeable form. The group's experiments in connecting disparate bodies, needs, and desires in temporal, contested networks of future-imaging invokes an art-historical frame that emphasizes process over the reified work-form, lifeworld over art world, anticipation over preservation. As such, Mobile Image is part of what art historian Gene Ray has called "avant-gardes as anti-capitalist vector," pointing and working beyond bourgeois notions of autonomy—whether of subject or object—toward what he calls "enactment," a type of agency that is "political without ceasing to be artistic or aesthetic."[5] This frame allows us to draw relationships between methods and models ranging from

Bertolt Brecht and the Bauhaus to contemporary utopian thinkers such as Donna Haraway and Oskar Negt and Alexander Kluge, not in linear terms but as a constellation, a set of coordinates that provide new orientations and potentials for transformation. Mobile Image activated a technics of aesthetics that was, and remains, an inspiring example and lens for contemporary communication media practices.

CRITICAL RECEPTION

In the introduction to his fiftieth anniversary reissue of *Expanded Cinema* (2020), Youngblood refers to Mobile Image as "by far the most important media art career over this half-century."[6] Yet, the group's work and its milieu remain woefully underexamined. In the 1970s and 1980s, that work was occasionally covered in brief newspaper and magazine reviews—mainly in the form of human-interest stories; it received negligible attention from the "art world" and its attendant publications. The lone exception was Youngblood, who published some significant essays in the mid-1980s and worked on an unpublished manuscript on the group, sections of which likely date from the late 1980s or early 1990s.[7] Since then, with few exceptions, art historians have taken little notice. (Arguably, the most important "reception" of the work during that time was by artists such as Ulysses Jenkins and Ben Caldwell, who, as will be discussed, were crucial to the development and facilitation of Mobile Image's work and subsequently built on it.)

More recently, with the rise of our hyperconnected culture, exhibitions and scholarly and critical texts have drawn some renewed attention to the group, mostly situating Mobile Image within histories of and discourses on interactive art, networked communication, the web, and social media. In his 2009 survey *Art and Electronic Media*, Edward A. Shanken devotes a brief paragraph to *Satellite Arts* and *Hole in Space*, in a subsection titled "Networks, Surveillance, Culture Jamming." Shanken contextualizes those works within a mid-1970s to early 1980s moment marked by an interest in interactivity and "the creative potential of telecommunications media." Although he identifies in the art of this moment an overall "Brechtian desire to wrest the power of representation from the control of corporate media and make it available to the public," Shanken does not address how and whether the works he covers would accomplish that. He describes Mobile Image in general terms, as art events for people to come together and connect in new ways. The group is similarly described in *Net Art Anthology*, an online archive (and accompanying catalog) assembled by the new media arts organization Rhizome. Launched in 2016 and completed in 2019 to accompany the New Museum's exhibition "The Art Happens Here: Net Art's Archival Poetics," the site is an important repository of photographic and video documentation, artist statements, and descriptions of works of art, including *Satellite Arts*, *Hole in Space*, and *Electronic Café*. The authors assert that Mobile Image "sought to critique corporate and government control of telecommunications and model the social and cultural possibilities of networks for the people," but without a sense of what exactly such a critique entails and how its politics might unfold.[8] The Rhizome project frames Mobile Image as part of a particular lineage, as belonging to "a possible net art canon."[9]

Annmarie Chandler's chapter on Mobile Image in the essay collection *At a Distance: Precursors to Art and Activism on the Internet* (2006) includes an extensive interview with Galloway and Rabinowitz, punctuated by relatively thorough descriptions of all three works, historical and biographical details, and thoughtful commentary by the author. Regarding *Satellite Arts*, for example, Chandler notes:

> This exploration of "being" and inhabiting virtual environments was made over a decade before philosophers would again become invested in concepts of "place" as distinct from the time/space/pace discourse dominating emergent communication technologies. More interestingly, *Satellite Arts* incorporated most of the elements now being argued as integral for the formation of identity in unexplored and uninhabited landscapes and for the act of implacement to occur. The social inquiry necessary for establishing "where I am, how I am together with others, and who we will become together" was stimulated through interactions between the real (the "here") and the virtual (the "there").[10]

Chandler's overall account is more nuanced and pointed than that of those who cast the work of Mobile Image as a general means of forging new connections or way to put advanced technologies in the hands of "the people." She describes their overall project as a "confrontation of the status quo for hijacking the social imagination from exploring th[e] potentials for real-time, multimedia collaboration," and *Electronic Café* as a response to the Orwellian "antiutopian vision of a totalitarian communication order." "With uncanny insight into the adverse social potentials of a gestating digital information society," she continues, "Mobile Image designed a project to create a model for [as the artists put it] 'a place where a globally networked culture might emerge that would enable consumers of information to evolve into the architects of services that served their interests rather than the interests of Microsoft.'"[11] Chandler concludes by suggesting, in retrospect, why Mobile Image has received so little attention, despite the continued relevance of their work and its underlying politics: "Galloway and Rabinowitz's work, however, was troubling to a powerful social reality in which 'celebrities dance' in the thin veils of civility between a demographic and marketable identity abridged to the 'grid of one.' Their emphasis, therefore, on inclusiveness, cultural diversity, and interdisciplinarity in their inspirational telecommunication art represents a heroic quest in animating an intervening social paradigm for the development of emergent electronic and digital communication environments."[12]

Others have emphasized the participatory aspects of the group's practice. The 2008–2009 exhibition "The Art of Participation: 1950 to Now," organized by the San Francisco Museum of Modern Art, featured a staging of *Hole in Space* in the form of two opposing screens showing video footage of the original 1980 event from each location.[13] Curator Rudolf Freiling explains that the works in the show belong to "a lineage of artistic approaches that include communication art, institutional critique, relational aesthetics, and social practice" and, more broadly, to a culture "that has fully embraced the new tools of social networking."[14] The catalog celebrates Galloway and Rabinowitz for having "exploited satellite technology as a dynamic medium for live performance and social interaction."[15] In a review of the exhibition, however, Tom McDonough singles out *Hole in Space* as an example of superficially interactive art that uncritically embraces now-dominant "egocasting technology." Unlike Chandler's account, his critique seems based more on the two-channel video documentation at the San Francisco Museum of Modern Art—and the museum's largely apolitical framing of it—than the actual work, its creators, and the complexity of encounters that it contained: "Geographic distance may well be eliminated, but psychic and social distance remain stubbornly intact. In both New York and Los Angeles, the television images appeared behind large windows, and even in the video documentation, we can feel those sheets of glass separating the two groups, each of which looks a bit like a pack of caged animals. Interaction is simultaneously summoned and forbidden."[16] For McDonough, Mobile Image "occupies a conceptual crux, a turning point between vanguard practices that often sought to demystify the normative categories of artmaking—the object, the artist, the viewer—as part of a larger critical program aimed at producing new, nonhierarchical social relations, and practices that have absorbed technological advances only to affirm our contemporary economy of spectacle and commodification." In contrast, he praises works that employ a "strategy of counternarcissism" by deliberately rupturing participants' experiences and generating critical reflexiveness—something he sees as absent from *Hole in Space*.

The 2011–2012 exhibition "Collaboration Labs: Southern California Artists and the Artist Space Movement" at the 18th Street Art Center (part of the landmark "Pacific Standard Time: Art in LA 1945–1980" series of exhibitions) was the most prominent and extensive presentation of Mobile Image, one of five artists/groups featured in the show.[17] It featured archival videos, photographs, printouts, and preparatory drawings from *Satellite Arts*, *Hole in Space*, and *Electronic Café*. In his catalog essay, the show's curator, Alex Donis, situates Galloway and Rabinowitz within the context of collectivism and alternative spaces and networks in Los Angeles–based "media art" from the 1960s through the 1980s—what he refers to as "LA's illusive communal art terrain"—something we address, mainly in our chapter on *Electronic Café*. For Donis, Mobile Image belonged to "an artistic and ideological movement that forever altered the artist/spectator/participant dynamic in the corpus of West Coast art practice from the middle of the 20th century to the present." Like others, he sees their work as deeply resonant today: "As we text and tweet our way into the 21st century, Rabinowitz and Galloway's prescient strategies in public communication remain at the forefront of global connectivity—exemplars of communal curatorial praxis."[18]

In that same catalog, Julia Bryan-Wilson builds upon Donis's overview, providing a brief but insightfully nuanced framing of the works in the exhibition—with far-reaching implications. Bryan-Wilson places Mobile Image in a Southern California context of "technologically informed" artists pursuing a type of collectivism calibrated to counter the forces of mainstream, commercial media. She asks: "How did these collaborations envision electronic networks and new media as a means to forging new aesthetics? How did they use technology as a complex tool for political organizing within the social movements then emerging in Los Angeles, including feminism, gay liberation, and cross-racial community building?" Although Bryan-Wilson acknowledges that fully answering such questions is beyond the scope of her catalog essay, she lays out a productive methodology for doing so based on the particularities of that local context. Crucially, she rejects the "overtly simplistic binary [of] collusion [versus] subversion," which generally obscures distinctions between different places and times and, more to the point, "does not do justice to the complicated ways that artists in Los Angeles have historically negotiated their relationships with technology." Where McDonough invokes a monolithic economy of technological spectacle and commercialism, to which works of art either stand in opposition or succumb, Bryan-Wilson recognizes that artists in this context had a more ambivalent approach to art/tech collaboration, working both within and against the apparatus.[19] Los Angeles in the 1970s and 1980s was, she explains, "an incubator for alternative modes of making that take advantage of proximity to the resources, publicity machines, skilled Labor, and cast-offs of mainstream media." The artists under discussion took up these tools to create "gathering places, pockets of collaborative activity, and loose networks of affinity" in a city too often caricaturized as decentered and placeless.[20] Rather than wholly reject emergent technologies because of their commercial origins and spectacular uses, artists such as Mobile Image *mobilized* them, "seeking to utilize, interrupt, or transform the means of production themselves through their commitment to counter-hegemonic models of making."[21] Our study takes up, at least in part, Bryan-Wilson's approach.

Kris Paulsen's 2017 book, *Here/There: Telepresence, Touch, and Art at the Interface*, includes part of a chapter on the work of Mobile Image, comprising the most in-depth examination of the group's work to date.

Paulsen's overall focus is on art in which screens function as spaces for not only telecommunication but also embodied relations via remote, mediated, sensory interaction. "Telepresence and its tactile interventions in and through screen space," she argues, "complicate the boundaries of our bodies, extend our corporeal agency and influence, and blur the distinctions between physicality and virtuality." As a result, the screens employed by such artworks "place the viewer in specific social and political as well as somatic relationships to what they show," eliciting critical reflection on the spatial and temporal deceptions of mainstream media.[22] Paulsen resists the simplistic notion of electronic media as "immaterial" and therefore devoid of indexical trace and the physical and existential implications—and complications—of that trace. She thus places Mobile Image in a lineage of work that treats the screen as a space for "telepresent" communication, starting with specific examples of late 1960s and early 1970s video art and extending to twenty-first-century artworks about drones. *Satellite Arts* is the main focus of the chapter, specifically its transformation of the television screen into a "place" for reflexively embodied interaction, for immaterial touching. Contextualizing it within a popular discourse on telecommunicative "touch," including other satellite-based art and contemporaneous satellite TV spectacles, Paulsen argues that the work resisted what Rosalind Krauss famously described as video art's "aesthetic of narcissism" by establishing a phenomenological relationship between self and other in the screen: "By combining the two video feeds into one and synesthetically collapsing vision and touch, *Satellite Arts* held in suspension the binaries that structure embodied existence—here/there, now/then, self/other, real/virtual—and actualized Maurice Merleau-Ponty's notion of the 'chiasm,' a condition of simultaneity that is only ever eminent in the physical body. Galloway and Rabinowitz used satellite technology and 'real-time' video images to hypothesize an ethics of engagement with others in mediated environments and to model a phenomenology of telepresence."[23] Paulsen's nuanced account relates to Chandler's in its discussion of identity as a dialectic between here and there, real and virtual, and to Bryan-Wilson's in that it establishes a technological ambivalence at the core of Mobile Image, which at once embraced government and military technologies (NASA-funded satellites) and the liberating possibility of "immateriality" and elicited reflection on the ethics and politics of such use: "Galloway and Rabinowitz reimagined the relationships a viewer could have to herself and to others by using a mediated image, and consequently they provoked the viewer to rethink how she might be responsible for and how she relates to the images she sees on the television screen. They demonstrated how combining the here and now with the there and then might stage a chiasmic experience in video space. One cannot simultaneously experience being both subject and object in the physical, material world, but the 'image as place' diagrams what this phenomenological experience might be like in the 'immaterial' world."[24]

LABORATORY

In the early 1930s, Brecht proposed a similarly expansive technics of aesthetics. During the rise of fascism, in search of a way to resist and provide a forceful alternative to the cultural industry mobilized by the Nazis, he pondered the conventional function of art, its spaces, and its audiences. More generally, he questioned the role of the artist, who had seemingly exclusive access to the tools of creative expression and whom Brecht accused of supplying a merely symbolic liberation from ideology. In 1932, he famously demanded the *Umfunktionierung*, or reappropriation, of the radio apparatus, observing that "the radio is one-sided where it should be two-."[25] According to Brecht, the radio should be changed from a medium of distribution, of "mere sharing out," to one of actual communication: he envisioned a "network of pipes" that would allow the audience to transmit as well as receive. By turning listeners into producers, such a two-way radio would transform the dynamics of social networks. It would, he argued, "bring [listeners] into a relationship instead of isolating [them]." Rather than create an alternative sphere of subjectivity and communicative activity or replace one set of subjects with another, this change in relationality aims to convert the system of individual (aesthetic) consumption into one of social production of public attitudes and perspectives. Crucially, this transformation of the apparatus has to be innovative both technically and ideologically: "It must follow the prime objective of turning the audience not only into pupils but into teachers." The entire landscape of communication and knowledge production, the role of each individual within this network of subjects, would be drastically altered, at once providing new capabilities and demanding a new form and conception of subjecthood itself.

Mobile Image operated amid the emergent institutionalization of another technological apparatus, one that would soon become dominant. Adopting a similar notion of *Umfunktionierung*, the group's elaborate projects connected people and peoples across geographical, ethnic, racial, gender, and class divides. These projects at once prophesied, embraced, and challenged the notion of a hyperconnected world, enabling participants to recognize their own place in the apparatus, experiment with the technologies that would bring about that world, and consider the politics, histories, and relations that pervade techno-social development.

Galloway and Rabinowitz shared Brecht's caution not to mistake technical novelty for political and *geistige* (intellectual/spiritual) progress, not to submit to the "renovation [of] ideological institutions on the basis of the existing social order by means of innovation."[26] In an unpublished 1977 text that laid the foundation for *Satellite Arts*, the artists wrote: "Our technological society is in an era of transitional mutation from an industrial age to an electronic age. It is an age where dominant agencies of communication produce the message systems that cultivate the dominant social patterns. They structure the public agenda of existence, priorities, values and relationships. It is these mutating consequences that become forces of acculturation, creating a new symbolic order."[27] To Brecht, technical innovation, conventionally conceived, was in danger of primarily expanding the distribution of content and thus economic and ideological domination under the guise of progress. He lambasted "the colossal triumph of technology at last to be able to make accessible to the entire world a Viennese waltz and a kitchen recipe."[28] Mobile Image echoed these concerns: "At present, the cultural benefits of satellite technology, to the larger public, come packaged in the form of human competition such as boxing and the Olympic games."[29] Interestingly enough, the artists not only criticized the lack of "educational" quality of broadcast media but also pointed to what they thought of as an emerging socioeconomic pattern of increasingly competitive individualized and individualiz*ing* attitudes, both of which they saw as contributing to an isolation of audiences rather than an engaged relationality between and among them. It is important to point out that within this logic, an *educational* approach to technology would mean the establishment of a communicative, two-sided relationship of meaning and knowledge production, not simply the transmission of information.

These reciprocal interrelations, resulting in the transformation of the very being or condition of subjects and objects, the *de-reification* of people and things as well as the processes that connect them, do not lend themselves to leaving an art-historical footprint. The avant-garde vector entails a commitment to questioning the autonomy of the (art) object and of the subject that makes and uses it. For the people to become "new men and women," to share ownership of the devices and apparatuses with which needs are addressed and new images, imaginaries, and conditions are created, for the public to partake in the shaping and overhauling of the given in the building of a new society, the products made and disseminated would have to comprise more than symbolic forms of participation. They would have to be tools that actually grant agency. In 1922, László Moholy-Nagy distinguished between production and reproduction as follows: "Creative activities are useful only if they produce new, so far unknown relations."[30] *Re*production is defined as the "reiteration of already existing relations." He continued: "Since it is primarily production (productive creation) that serves human construction, we must strive to turn the apparatus (instruments) used so far for only reproductive purposes into ones that can be used for productive purposes as well. This calls for profound examination of the following questions: What is this apparatus (instrument) good for? What is the essence of its function? Are we able, and if so to what end, to extend the apparatus's use so that it can serve production as well?"[31] Moholy-Nagy refers to both the "functional apparatuses" of which the human being is composed (the "cells as well as the most sophisticated organs") *and* the prosthetic devices used to extend further into and interact with the world (the phonograph, photography, film).[32] Technology is inherently neither progressive nor oppressive; its function depends on how it is utilized with regard to human potential and how selfhood relates to its environment. Being more human, or rather less alienated in modernity, does not mean reverting to a less technological, less materialist stage. It means making and using things differently.

This approach was pointedly refined and articulated by Russian Constructivist theorist Boris Arvatov, who, in a 1925 essay titled "Everyday Life and the Culture of the Thing," called for the production of "socialist objects." To Arvatov, the avant-garde's task was to rethink and redefine the conventional relationship between subject and product, to make a thing that would function not as a commodity but "like a comrade." As Christina Kiaer points out, whereas Marx sought to substitute a truer, interhuman relationship for the "reified relations between people and social relations between things," Arvatov insisted on the production of a new object (rather than its abandonment altogether).[33] The system of production is thus understood not as an alternative or parallel dematerialized sphere but rather as a materialist one, in the sense that it continuously struggles over the reenactment of conventional forms of and attitudes toward ownership in labor and consumption. It is intricately connected to the material conditions of everyday experience and the technological means that create it.

At a crucial moment in the development of what is now often called the "Digital Era," Galloway and Rabinowitz recognized that then-emergent telecommunication and network technologies could be made to function in this way, as "socialist objects," as comrade things, and that, given the increasingly rapid advancement of such technologies, this potential was sure to grow exponentially over the coming decades. Yet, they were also certain that, despite the promises of prognosticators caught up in the techno-euphoria of the time, this potential would not automatically fulfill itself via the mass dissemination and adoption of novel devices.

Echoing Moholy-Nagy's sentiment, Mobile Image declared: "We don't produce artifacts, we produce living events that take place over a period of time, to facilitate a quality of human to human interaction."[34] The artists repeatedly referred to their work as a social and political laboratory and model, as creating opportunities for explorations in communication and connectivity that would mobilize latent potentials and yet-to-be discovered capabilities of existing technologies, as well as transform them according to present and future needs. As they stated in their foundational notes, their goal was to produce "PROTOTYPE COMMUNIT[IES] and provide EMERGING TECHNOLOGICAL ENVIRONMENTS."[35] Or, as they put it elsewhere: "Somebody has to be creating models to liberate people's imaginations so they can apply them to *hope* and the *possibility* of redefining these technologies."[36]

Related themes—the mediation of bodies, spaces, and subjectivities; public/private dialectics; utopian imagination—weave through the group's work, binding these projects to broader contexts and discourses while raising fundamental questions: How do certain technologies facilitate new forms of public life? Who gets to participate? What is the *function* of art? How do such technologies impact what is imaginable in the first place? These are economic questions in the expanded sense, including both material and immaterial production and distribution: What do we, as a society, make? How and for whom? They are questions of relationality, of power and access, of owning labor and selfhood, representations and voices, histories and narratives.

That relationality, the ability and opportunity to perceive, reimagine, and reorganize, as a social subject, a set of given and potential structures of interaction and interdependence, was paramount in Mobile Image's approach. It is what allows their work to be understood as a laboratory for the production of and experimentation with so-called socialist objects. Relational concerns pervade the abundance of notes, plans, and texts in their archive. For example, an early 1980s collection of inquiries and self-imposed guidelines contains the following questions and directives:

> Who is the audience, what is the public dividend, what is the product being sold and to whom?
> Transform individual competence and social effectiveness?
> Distinguish between what we CAN do and what we CHOOSE to do, individually and socially
> The real issue is that of allowing access to the whole picture of what is going on
> Meta-technology: We feed off of ideas/ideas are a sort of technology/ideas require a technology
> Broadcast → Communication/quality of the human communicative experience . . . /new social spaces[37]

The vision and practice of Mobile Image presents a challenge to art histories forged during the Cold War, when scholars distrusted or outright rejected avant-garde models of artistic production rooted in socialism as a viable political option, in "scientific thinking" as a critical-artistic practice, and in the validity of creating a Leftist popular culture.[38] We propose that Mobile Image contributes to a different narrative or methodology—one that redefines understandings of both art and technology and hence of the relationship between artists and audience, a method that truly innovates political transformation rather than renovating binaries of aestheticized imaginations versus material determinisms.

FANTASY

Today, the Mobile Image archive contains, along with documents related to the group's work, an abundance of fascinating relics of modern technological aspiration, promise, and potential: ninetieth-century maps of transatlantic telegraph cables, boxes of newspaper clippings celebrating all sorts of transportation and transmission innovation, Sputnik models, toy ray guns, pulp sci-fi novels, a bobblehead of Nikola Tesla, green toy soldiers with communication field packs, Jetsons space gadgets, and much more. Collected since childhood and prominently displayed throughout the archive and in Galloway's workspace (Rabinowitz passed away in 2013), these artifacts are reminders of the dreams and visions that propelled Mobile Image's projects forward. Although certainly bound by gendered, racial, and class-based categories that made such dreams more accessible to some than others, they accentuate an approach to artmaking that sought not only to reflect reality critically in all its facets but also to challenge its institutions and transform the very notions of production, function, site, and audience. Art was to have *usefulness* in the making of new worlds. Fantasy was to be its engine.

Galloway and Rabinowitz were, however, keenly aware of the pitfalls of fantasy within a capitalist system that offers prepackaged and commodified dreams and visions and relies on commercialized myths of technology's inherent progress. In a 1987 conversation with the artists for the journal *High Performance*, Gene Youngblood put it this way: "We've been talking about the communications revolution. People take this in one of two ways: either it's some kind of '60s, hippie, utopian idealism or it's a marketing scam by industry, you know: 'The communications revolution is here, and our product . . .' There has been no middle ground discourse between those two extremes. I would just like to point out that any interesting thing that people like Kit and Sherrie would do with this technology would by definition have to be *a model of what a communications revolution would be like if there were one*."[39] In Mobile Image's practice, the utopian-fantastic dimension was to manifest as *process*, pursued in the laboratory of their work rather than as a delivered component of the work's technical materiality. Although it would apply differently to different groups and individuals, and would certainly not be linear or universal, progress meant actively engaging in the making of the future through a reconfiguration of a hegemonic techno-logic and the tools that have been wielded according to that logic in the production of the present and past.

Crucial then, as now, was a question recently summarized by Peter Thompson: "After the apparent death of the grand narrative of progress," is it possible that "hope can still exist in anything other than an atomized, desocialized, and privatized form"?[40] Are dreams of a better world possible not as commodities offered for individual consumption but rather as the focus of collective and collaborative action? During the late 1960s and 1970s, artists, activists, and theorists had been cautiously optimistic that new developments in telecommunications technology would enable such dreams and action. In response to the proliferation of TV stations and in-home receivers, of satellite transmission, cable distribution, and video equipment, "guerilla television" sought to wrest broadcasting from the military-industrial complex. Conferences and exhibitions, meanwhile, pondered "TV as art." Writers including Allison Simmons, Ivan Illich, Herbert Schiller, and Raymond Williams cast a critical eye on mediums and the media, aiming to separate technical-transformative potentials from commercial-ideological applications.[41] According to Schiller, the desire for greater access to and flow of information was both a legitimate emancipatory aspiration and an imperialist strategy. He called for the development of a new "communications consciousness" for an understanding and treatment of all "cultural-communications technologies" as economies, hence as contested sites of the "production and distribution of communications-cultural outputs," as a crucial part of a struggle over power.[42] Nonetheless, progressivist myths of liberation via technological advancement only became more ingrained in the American psyche during this time. Those myths would be further amplified by the mass production and marketing of personal computers in the late 1970s and early 1980s. Hewlett-Packard, Commodore, and Tandy Radio Shack advertised their PCs as affordable and unobtrusive machines that would aid the average family and small business to organize their lives, safely connect with the world, and have fun while doing so. Going one step further, IBM and Apple famously tapped into Cold War fears of technocratic homogenization, presenting their products as democratizing machines, as creative and individualized solutions to the threats of mass-produced experiences and lifestyles.

Some, including Mobile Image, were highly suspicious of the built-in promises of the ever new. And yet, there was increasing recognition that the 1960s New Left notion of a monolithically oppressive mass media apparatus and culture/consciousness industry was too simplistic. Critics and philosophers such as Youngblood, Williams, Hans Magnus Enzensberger, and Oskar Negt contributed to an emergent critical theory of new media practice and relationality.[43] In his influential 1970 essay "Constituents of a Theory of the Media," Enzensberger explains: "The New Left of the sixties has reduced the development of the media to a single concept—that of manipulation. This concept was originally extremely useful for heuristic purposes and has made possible a great many individual analytic investigations; but it now threatens to degenerate into a mere slogan that conceals more than it is able to illuminate, and therefore itself requires analysis."[44] To Enzensberger and others, there is no "pure" communication, knowledge, or experience; they are always manipulated, always mediated. As tools, communication devices have no default function but rather are subject to power and access, to the ideologies that frame them, that limit them to predetermined capacities, or that open them to latent forms of utility: "It is wrong to regard media equipment as mere means of consumption. It is always, in principle, also means of production and, indeed, since it is in the hands of the masses, socialized means of production. The contradiction between producers and consumers is not inherent in the electronic media; on the contrary, it has to be artificially reinforced by economic and administrative measures."[45] Different understandings of the media could reveal new ways of using them. The challenge was how to use new technologies in ways that did not merely perpetuate entrenched myths of either liberation or domination.

Mobile Image's notion of art as a laboratory for future-imagining and making based on the transformative potential of social labor, relationality, and the specific hopes and needs of users recalls Ernst Bloch's concept of "concrete utopia" or utopia as method. According to Bloch, utopia is a process of confronting an incomplete state of existence, where "too much is full of something that is missing."[46] Hope is the longing to fill in and shape this missing portion, not in some mystical sense or according to a preordained destiny but rather as a dialectic between human and material forces, an active construction of what history has repressed, forgotten, and left unfulfilled. In *The Principle of Hope*, Bloch observes that the human being "is not solid," and asks "to think oneself into what is better."[47] To be knowingly "not solid," incomplete, becoming—or "fermenting," as he evocatively puts it—can be liberating, for it at once opens space for "dreams [to] drift in" and causes people to be "much more intensely present."[48] When art is conceived as a utopian process in this sense, it can not only play an active role in the production of society but also potentially construct a *different* society. And technology—itself a material and intellectual prosthetic, an augmentation of the knowingly incomplete human being—can provide the tools for exploration, experimentation, and realization within such a process. To Bloch, the "motor" of true progress is "the real possibility of the not-yet-conscious, the not-yet-become" at a historical moment that contains the material circumstances for that becoming.[49] At just such a historical moment, Mobile Image aimed to be that motor.

To do so, the group critically made use of the productive processes of fantasy, as defined by Oskar Negt and Alexander Kluge in their groundbreaking 1972 work *Public Sphere and Experience: Toward an Analysis of the Bourgeois and Proletarian Public Sphere*. As understood here, fantasy is not just what can be pictured or projected outside of given norms and limits but also a future built on the ambiguities and fissures of existing and latent ideas, histories, and circumstances, confined neither to an all-prescriptive progressivist public ideal nor to discrete and disconnected individuated dreams. It is the potentially unassimilable alienation produced by the overlaps and incompatibilities between the general social horizon of experience and the everyday life experience of individuals and particular social formations. As artists, Mobile Image sought to turn the tension between the given and the possible into the motor that is the fantastic imagination. The awareness of telecommunications technologies achieved by "seeing oneself seeing" was only part of this critical aesthetic process; beyond that lay the expansion of artistic labor, of art's institutional apparatuses, of what Bloch calls the "still undisclosed."[50]

COUNTERPUBLICS

The material and ideological reorientations and reorganizations modeled by *Satellite Arts*, *Hole in Space*, and *Electronic Café* would contest the increasing privatization of experience through individual consumption of and through technical gadgets while also critically transforming the very concept of the public sphere as a site for social labor, for the collective making of ideas and meaning. Among the issues they explicitly pursued, the artists listed the following:

> What constitutes the public interest?
> What will govern [its] growth and development?
> Who is the audience, what is the public dividend, what is the product being sold and to whom?
> Politics of ideas/owning ideas/selling ideas[51]

Ultimately, Mobile Image hoped to facilitate what Negt and Kluge call "counterpublics." Implicit in this concept is an understanding of the public sphere and its apparatuses of publicity as not excluding the private but rather organizing it as part of an overall experience, at once socially binding and replete with contradictions between ideals, expectations, and everyday life. The public sphere is a site of struggle as much over psychological, emotional, and ideological territory and technologies as over material space and tools. Information and communication, visibility and representation are key. As Negt and Kluge explain: "The public sphere denotes specific institutions, agencies, practices (e.g., those connected with law enforcement, the press, public opinion, the public, public sphere work, streets, and public squares); however, it is also a general social horizon of experience in which everything that is actually ostensibly relevant for all members of society is integrated. Understood in this sense the public sphere is a matter for a handful of professionals (e.g., politicians, editors, union officials), on the one hand, but on the other, it is something that concerns everyone and realizes itself only in people's minds, in a dimension of their consciousness."[52]

In reaction to the revolutionary spirit of the student, antiwar, and civil rights movements of the 1960s, an emergent conservatism seemed to be exacerbating the erosion of the so-called bourgeois public sphere. According to philosopher and sociologist Jürgen Habermas, this sphere had once functioned as an arena for critical exchange among autonomous citizens, a site of productive dissent, set to "mediate between a society of private property owners and the state."[53] Culture—the visual arts, literary criticism, talk shows, and so on—was the primary tool of such activity, and matters addressed were meant to be universal based on the needs and interests of society as a whole. As Habermas argues, the erosion of this arena was rooted in the advent of nineteenth- and twentieth-century capitalist democracy, under which the channels of critical exchange were increasingly commodified and the state became the protector of the general population, a guarantor of public welfare in a system of free-market capitalism. The bourgeois public sphere ultimately turned into what Habermas called a "façade of legitimization." It veiled the production of social hegemony, while offering up ideals of the greater good in place of those of discord, dissent, and debate. By the mid-1960s in Europe and, a bit later, in the United States, the discourse around the public sphere was increasingly framed by questions of class, of ownership of the means to make and disseminate ideas and opinions, images, and representations. By the 1980s, the backlash against the diversification of American culture during the previous decades appeared to have begotten a public sphere in which social norms and values were affirmed and cemented rather than subjected to ongoing renegotiation.

In response, certain art activists and social advocates challenged the very idea of the public sphere as a site of harmony and homogeneity.

Recognizing that such a sphere had always been exclusive, always limited by class, gender, race, and ethnicity, they demanded the inclusion of a plurality of experiences rather than the prescription of universal moral standards. This call for inclusivity offered a powerful critique of the bourgeois public sphere's normative capacity, as described (and ultimately advocated) by Habermas. Rather than assimilate a multiplicity of histories and imaginations into a general cause or seek to complement or complicate the existing public sphere, "counterpublic" art questions the underlying dynamic of what is deemed "public" and "private." Refuting that binary, it aims to articulate the relationality—a central term in Negt and Kluge's analysis—between the spheres where experiences are produced and circulated, and to posit such relationality as an experience in and of itself.[54] As Miriam Hansen points out, this counterpublic sphere of experience presents a radically new definition of what it means to participate in a social structure: "While Habermas' notion of public life is predicated on formal conditions of communications (free association, equal participation, deliberation, polite argument), Negt and Kluge emphasize questions of constituency, concrete needs, interests, conflicts, protest, and power."[55] The counterpublic sphere is based on a pragmatic notion of commonality that redistributes the tools of cultural production according to temporary alliances formed around problems and experiences, including those of exclusion and division.

It is therefore not enough just to make technology more "public" by expanding audiences and access, by offering more opportunities for participation and self-expression. To be truly communicative, it must facilitate productive relations between the private and the public, conceived as dynamic, multilayered, and intertwined rather than autonomous and monolithic. Accordingly, a counterpublic art practice would be based on a politics of perception and the recognition of the subject (and self) as a locus of entwined "public" and "private," social and individual, forces. Technology and media—and this applies to all, not just "new," media—would thus be seen as always inscribed with history, meaning, and predetermined function. Communication would entail a relationality between public, art, and technology and how they interrelate as entangled sites of struggle.

Galloway and Rabinowitz attempted to model what this could look like in practice. They understood themselves as facilitators and Mobile

Image as a vibrant, heterogeneous, unpredictable, and often contradictory but always evolving assembly of producers. To convene such an assembly in a counterpublic, fantastic way would, they hoped, yield neither what has been conventionally considered a public or community nor what has, according to the public sphere's dichotomous logic, been understood as its opposite.[56] Mobile Image was determined to remain highly self-conscious and critical toward their own place within the technological processes they facilitated, as well as the traditional function of art with regard to its makers and audience. What is the role of the artist in the economy of aesthetic production? For whom is something being made and to what ends? How is the audience and its function defined when the purpose of the work is not disinterested contemplation, spectatorship, or consumption but rather an ongoing collaborative navigation of old and new representations, attitudes, and experiences, the examination and transformation of relations? As the artists themselves put it: "Who should define standards, conduct tests, analyze results and make them known?"[57]

Approaching the public sphere as a site of struggle over how and by whom certain experiences are limited and crippled, made possible, visible, and valuable, Mobile Image did not posit a predetermined alternative to or commentary on the current order. Rather, as stated in the artists' notes, they strove to attain:

> Flexibility and dexterity in the social infrastructure
> EDUCATIONAL ENCOUNTER
> Creative conversations/creative solutions/experience sharing.[58]

"Experience" is a central term in Negt and Kluge's definition of a "proletarian" counterpublic, and "experience sharing" as an "educational encounter" evokes the Brechtian idea of everyone being both learner and teacher—key to the emancipatory potential of experience. When Negt and Kluge use the term "proletarian," they intend it in an expanded sense, beyond the singular notion of proletariat as a form of existence or identity. It is understood as an attribute of existence and experience, marked by being separated from the means of production, not only industrial production but "all productive capacities," including the resources and tools to make and shape one's ideas, perspectives, knowledge, and history.[59] Proletarian experience is thus largely characterized by the alienation between, on the one hand, the "general horizon of social experience"—defined by institutional narratives and blueprints, official accounts of being and seeing, past, present, and future—and, on the other, the everyday impressions, encounters, and experiences that may or may not fit into what that "horizon" deems valid, important, factual, possible. "Real experience," Negt and Kluge explain, "is torn into two parts that are, in class terms, opposed to one another."[60] If it can be organized accordingly, this alienation, this unassimilable proletarian experience, carries the potential for change.[61] Mobile Image modeled functional, technological processes for harnessing and reorganizing that experience.

SITUATION

As part of the evolution of the "new social movements" in the 1970s and 1980s and postmodernist revisions of authorship and hegemonic "metanarratives," several artists and writers in the United States similarly grappled with the task of redistributing and repossessing experience and its technologies.[62] Questions of individual and social identity were cast in terms of ownership of the categories and constellations of subjectivity and selfhood, and of the tools used to construct them and, ultimately, to be deployed in a process of fantastic innovation. How was one to make conscious and palpable an experience that would critically engage the stereotypes and naturalized outlooks and behaviors officially condoned by and in the public sphere without replicating and reproducing the logics and technics of self and other, center and margin, private and public? Postcolonial studies scholar Homi Bhabha committed to promoting a "politics of identity" over "identity politics," offering the following remarks about "how newness enters the world":

What is at issue is the performative nature of differential identities: the regulation and negotiation of those spaces that are continually, *contingently*, "opening out," remaking the boundaries, exposing the limits of any claim to a singular or autonomous sign of difference—be it class, gender or race. Such assignations of social differences—where difference is neither One nor the Other but *something else besides, in-between*—find their agency in the form of the "future" where the past is not originary, where the present is not simply transitory. It is, if I may stretch the point, an interstitial future, that emerges *in-between* the claims of the past and the needs of the present.[63]

To the author and activist bell hooks, writing in 1984, this "interstitial future" emerged at the site of marginality. What Negt and Kluge call "real experience" is, for hooks, the experience of selfhood as a contested relation between public and private, providing a revolutionary consciousness through a critical "sense of wholeness":

To be in the margin is to be part of the whole but outside the main body . . . Living as we did—on the edge—we developed a particular way of seeing reality. We looked both from the outside in and from the inside out. We focused our attention on the center as well as the margin. We understood both. This mode of seeing reminded us of the existence of a whole universe, a main body made up of both margin and center. Our survival depended on an ongoing public awareness of the separation between margin and center and an ongoing private acknowledgment that we were a necessary, vital part of that whole.[64]

The work of Mobile Image addressed questions of identity, or rather it openly struggled with its terms and signs of difference. Their practice should not be misread as an insipid type of "multiculturalism"; quite the contrary. The work was problematic in the sense that it at once complicated singular categories of identity and the apparatus that produces and maintains them, asking its participants to engage directly in the ongoing contest that is the imaginary and material construction of difference, and enacted those very categories, making them sense-able, at times reaffirming and at others transgressing them. The work as laboratory facilitated the counterpublic experience of ambiguity and contradiction without ignoring the very real manifestations of divisively oppressive identity markers. As the artist Daniel Martinez, a participant in *Electronic Café*, acknowledges: "[Mobile Image] realized the city was not a model. It was built by a lot of people with a lot of conflicting histories."[65]

That Galloway and Rabinowitz were in a position to accomplish what they did—and at such a grand scale—cannot, of course, be separated from their privileged status as white, middle-class, heteronormative artists. They had access to institutions, tools, people, and spaces not available to all. Conscious of this, they deliberately engaged various "marginal" communities, as will be discussed at length throughout this book. Yet, they never explicitly questioned their own position and the uneven accessibility to those components and to the "art world" itself, however conceived. Galloway and Rabinowitz often sought to elide that positionality, indeed their very presence, turning over the technologies and spaces at hand to participants who would not ordinarily have access. Such elision is itself a privilege though; others cannot simply disappear at will. And even when their "laboratory" explicitly foregrounded race, class, ethnicity, and other identity categories—as in *Electronic Café*—there was a lack of acknowledgment of the disparate levels of opportunity among the groups gathered. They were not all equally marginal.

For the artists and their collaborators, the core challenge was to figure out how to mobilize the various manifestations and images such that the sense of seeing described by hooks could facilitate a structure of allegiant formations between individual and social bodies. How could they productively communicate, share experiences, and find creative solutions without re-creating either the violence of a prescriptive common denominator or the relativist cacophony of unorganized idiosyncrasy? According to feminist scholar Donna Haraway, the first step is to acknowledge, make conscious, and celebrate that there are no "innocent" or "objective" positions in the patriarchal sense. All perspectives are mediated, all knowledge is partial; as she put it in 1988, "Feminist objectivity means quite simply *situated knowledges*."[66] Such knowledges are not relativist; rather, they exist in and forge contingent constellations: "The alternative to relativism is partial, locatable, critical knowledges sustaining the possibility of webs of connections called solidarity in politics and shared conversations in epistemology. Relativism is a way of being nowhere while claiming to be everywhere equally. The 'equality' of positioning is a denial of responsibility and critical inquiry."[67] Designed to be situated in this sense, the "prototype communities" of Mobile Image comprised proletarian bodies, subjects, and selves empowered to construct individual and social formations rooted in the latent potentials of the past and present.

These new relations, along with the politics of creating them and the desire to do so, were steeped in a technics of aesthetics, in which technologies are imaginatively and usefully refunctioned on behalf of a relationality between people, things, and contexts, experiences, tools, and needs. Whether it be the innovative mediation and navigation of bodies in televisual space (as in *Satellite Arts*), the joyful amplification and stress test of the urban public sphere (as in *Hole in Space*), or the production of fully fledged counterpublics through existing and new constellations of experience, alienation, and desires (as in *Electronic Café*), Mobile Image's practice in many ways resembled what critic and theorist José Esteban Muñoz terms "queer futurity," in which utopia exists in a future state beyond given confines and definitions of bodies, identities, and attitudes: "Queerness is not here yet. Queerness is an ideality . . . Queerness is a structuring and educated mode of desiring that allows us to see and feel beyond the quagmire of the present . . . Queerness is also performative because it is not simply a being but a doing for and toward the future."[68] To mention Muñoz here is not to designate Mobile Image as queer artists or their work as queer art but, in the making of new, fantastic subjectivities, to emphasize again the process of the avant-garde as vector over the reified form, the departure from existing constraints of history as understood as "fixed" and therefore of *being* as "ontologically static," toward a perspective onto the given as potential *becoming*.[69] Like Mobile Image, Muñoz frames experience in terms of ownership, of access to the mechanisms that make the past and present and thus the future; like Negt and Kluge, the writer's politics of redistributing that which and *who* can be thought and performed or enacted, and *how*, are Marxist insofar as Marxism is a philosophy and practice of collectively thinking the possible. He cites Bloch: "Marxism, above all, was first to bring a concept of knowledge into the world that essentially refers to Becomeness, but to the tendency of what is coming up; thus for the first time it brings future into our conceptual and theoretical grasp. Such recognition of tendency is necessary to remember, and to open up the No-Longer-Conscious."[70] With Muñoz, Galloway and Rabinowitz shared a desire of "doing for and toward the future" by emphasizing process, trying out, actively and continuously forging the fantastic.

COLLABORATIVE RESEARCH

The mid to late 1970s was a complicated time for artists to model their work on the research laboratory and embrace emergent technologies as tools of progress. Mobile Image was formed during a moment of general disillusionment, following a swell of exuberance for experimental collaborations between art, science, and technology.[71] That exuberance had been keyed to Cold War discourse and history, including rapid technological advances, the explosion of consumer society and corporate capitalism, and the American economic system's attempt to outperform communism on its own ground.[72] Between the mid-1960s and the early 1970s, leading universities, corporations, and museums supplemented the notion of individualism as an antidote to totalitarian conformity with an emphasis on collaboration, communication, cooperation, and broad participation as democratic ideals. What John Beck and Ryan Bishop call a "military-industrial avant-garde" emerged, supported by interdisciplinary programs designed to conjoin American art and scientific, technological, and industrial innovation for the common good.[73] Crucial to these programs was an attempt to claim collectivism as an American value, even if it meant sidestepping or massaging out the political and ideological problems of such military-industrial collaboration. "Finding ways of bringing advanced art and cutting-edge technology together," Beck and Bishop explain, "was conceived of as a way of unleashing the creative capacities of artists, scientists, and engineers, of sparking as-yet unimagined inventions of form and function, and of initiating new modes of inquiry unfettered by conventional distinctions based on professional loyalties, prejudices, or habits of mind."[74] A recuperation of the historical avant-garde's collectivist ethos would hopefully allow an otherwise hierarchical, corporate environment to generate radical transformation, "not merely a new aesthetic but a new social order."[75]

This progressivist aspiration resonated in Southern California, which had grown into an international center of advanced technology and media, with aerospace and defense corporations alongside prominent science-oriented universities and institutions and the massive movie and television industry. As Donna Conwell and Glenn Philips explain, "An open-minded creative class of scientists, fabricators, and industry professionals were both prevalent and amenable to collaboration," and as a result, "the region increasingly came to be seen as ideally suited for the merger of art and technology."[76] In 1969, Los Angeles inaugurated its own branch of Experiments in Art and Technology (EAT), originally founded in New York in 1966. LA's EAT facilitated collaborations between artists, musicians, curators, and scientists from the California Institute of Technology (Cal Tech) and Jet Propulsion Laboratory (JPL). This resulted mostly in perceptual environments, performances, and spectacular multimedia events, involving technologies such as closed-circuit television, fiber optics, infrared cameras, and sonar and wireless FM transmitters.[77] A couple of years earlier, the Los Angeles County Museum of Art (LACMA) launched its "Art and Technology" (A&T) program, which culminated in a 1971 exhibition of the same name. Maurice Tuchman, director of the LACMA program, saw A&T as a continuation of the historical avant-garde's interest in collaborating with industry "to reform commercial industrial products, to create public monuments for a new society, to express fresh artistic ideas with the materials that only industry could provide."[78]

Enthusiasm for such activities quickly dissipated, however, in the wake of the era's political and social upheavals and the questioning of power structures and the institutions, experts, and technologies that enable them. As Beck and Bishop explain: "By the time the A&T exhibition opened in 1971, what utopian spirit there might have been in the original ambitions for the project were unable to withstand the Vietnam War, Richard Nixon, and the shooting of students at Kent State University. The idea that US corporations could plausibly collaborate with artists to create new worlds of social progress was now evidence of complicity and corruption—technology was the problem and not the solution."[79] Critics berated these art-and-technology programs for their collusion with corporations and government institutions; what had just recently been celebrated as prototypes of R&D cooperation were now dismissed as profoundly naive and misguided, if not abject failures.[80] The more general movement toward collaboration and participation in 1970s art was similarly met with suspicion when it involved technologies derived from and commonly associated with the military-industrial complex. Emerging neoliberal economic and political orders intensified this suspicion, raising questions about the role of artists in the development and popularization of what were increasingly understood to be technologies of power, control, and exploitation—whether they be devices of rapidly and vastly expanding mediation, observation, or transportation. Tuchman himself noted this in 1971, explaining that, whereas there had been very little moral or political resistance by artists to work in "temples of Capitalism" when LACMA's program was conceived, things were now very different: "I suspect that if Art and Technology were beginning now instead of in 1967, in a climate of increased polarization and organized determination to protest against the policies supported by so many American business interests and so violently opposed by much of the art community, many of the same artists would not have participated."[81]

The 2019 essay collection *Hybrid Practices: Art in Collaboration with Science and Technology in the Long 1960s* attempts to recuperate some of the social and political aims and legacies of those earlier collaborations, despite their apparent failures and alignments with American capitalist innovation, corporatism, and consumerism.[82] As Anne Collins Goodyear notes in that publication, the disillusionment of the early 1970s "did not disrupt the deeper and more important spread of a fundamentally new relationship between the creative practices of art, science, and technology," now manifest in twenty-first-century discourses around private industry, the academy, and art.[83] This relationship, according to Goodyear, consists of the widespread acceptance and normalization of artistic-technological collaborations and "a more pervasive appreciation of art itself as an intellectual practice, one of the key insights of early innovators in the art and technology movement of the 1960s."[84] Beck and Bishop salvage the liberal aspirations of the original art-and-technology programs while recognizing their shortcomings, not only contextualizing them within the broad ideology of the Cold War era but also historicizing their subsequent denunciation as part of a limited and overly pessimistic "discourse of failure surrounding the avant-garde" in the 1970s, 1980s, and beyond.[85] Conversely, they criticize recent revivals that uncritically recast these oft-maligned 1960s and early 1970s programs as purely illuminating prototypes: LACMA's 2013 Art+Technology Lab (A+T), the MIT's 2015 Center for Art, Science, and Technology (CAST), and the 2016 EAT Salon sponsored by Nokia Bell Labs effectively depoliticized their predecessors, denying their well-rehearsed criticisms and, no less crucially, their original collectivist motives. Like several of the authors included in *Hybrid Practices*, Beck and Bishop reemphasize this utopian spirit, rooted in the historical avant-garde, underacknowledged in the literature, and missing from these more recent art-and-technology iterations.[86]

Yet, by jumping from the early 1970s to the 2000s, they, like Goodyear and others, leapfrog artists such as Mobile Image, who, in the immediate aftermath of those original programs, carried forward certain aspects of them, including a rejection of the production of objects and the notion of individual artistic genius, an emphasis on collectivity and interdisciplinary research, and a belief in the possibilities of turning the spectator or passerby into active participant.[87] Belying the notion that the period was marked only by retreat, Mobile Image offered different, self-directed, and distinctly self-critical models for art and technology collaboration. By presenting that collaboration as struggle, involving artists, devices, technical operators, and users, the group worked to politicize the underlying assumptions and values of their precursors along with their own, while retaining the utopian potential of art as technological research. Its practice and history show that this kind of alternate model existed and, at least to some extent, remains.

That practice and history resonates in more recent theories of and proposals for collaborative artistic-technological research. In his contribution to *Hybrid Practices*, for example, science historian W. Patrick McCray asks what if we instead approach such collaboration "*as experiments* and adopted ideas from the history of science to treat them thus?"[88] This would mean shifting the focus from products to process; "success" would not depend on whether the art object or event produced was both technically innovative and aesthetically pleasing, a standard that has historically obscured the primacy of the collaboration itself, including the essential roles and relationships of the various participants and the experimental methods they shared. McCray proposes the concept of "techno-aesthetic communities"—hybrid groups of artists and technicians oriented around a particular project—as a new framework. Here, the collaborative process is kept central, technical experts are acknowledged partners rather than behind-the-scenes facilitators, and the experiments conducted are seen in terms of a range of potentials, including artistic, scientific, and entrepreneurial. McCray identifies several overlooked examples from the late 1960s and early 1970s that, in retrospect, would benefit from this framework, collaborations that allowed for mutual engagement between artists and engineers and whose achievements were arguably more commercial than artistic in the conventional sense.

Erin Manning and Brian Massumi provide a more comprehensive and nuanced theory of creative, research-based collaboration in their 2014 book *Thought in the Act: Passages in the Ecology of Experience*. Noting a renewed interest in interdisciplinary collaboration as a core element of today's "knowledge economy," they recognize that it is not the problems raised by such collaborations that are new but rather "the extent to which policies intended to facilitate collaboration across the divides have been prioritized in government cultural and academic policies and in university structures. The way this has been done has created real opportunities—but also highly troubling alignments with the neoliberal economy." Such collaboration now comes with language of that economy, including thinking about results in terms of "deliverables," of "strategic importance" produced for "stakeholders."[89] The "neoliberalization of research" also subsumes the arts, which are now valued for their innovative potential while being professionalized via structures of quality control and quantitative measures based on economic results: "The neoliberal idea is never far that artistic activity is most productive, and socially defensible, when it feeds into industry tie-ins helping fuel the 'creative economy.' Moves within the academy toward institutionalizing research-creation are inevitably implicated in a larger context where the dominant tendencies are toward capitalizing creative activity. In that context, research-creation makes economic sense as a kind of laboratory not only for knowledge-based product development but for the prototyping of new forms of collaborative activity expanding and diversifying the pool of immaterial labor."[90] Manning and Massumi advocate for artists not simply to take a critical stance, as if they stood outside the system and did not in any way benefit from it. Instead, they should "work in the thick of the tensions—creative, institutional, urban, economic—and build out from them," to strive toward what Manning and Massumi call "immanent" critique, that is, "to inhabit one's complicity and make it turn—in the sense in which butter 'turns' to curd . . . to try to do [one's] small part to curdle 'research-creation's' annexation to the neoliberal economy."[91] Artists should not merely "apply" science and technology in innovative ways but rather establish models and platforms for "a truly transdisciplinary exploration of new territories of practice" to constitute a mutual interpenetration of processes not just a linking of disparate ideas and products—a collaborative technique that is process oriented, with results that are not preprogrammed but rather "experimental, emergent effects of an ongoing process." "What is key," Manning and Massumi argue, "is less what ends are pre-envisioned—or any kind of subjective intentional structure—than how the initial conditions for unfolding are set. The emphasis shifts from programmatic structure to catalytic event conditioning."[92] Such events should be organized with constraints; instead of the open-ended, free-flowing modes often associated with invention and improvisation and "creativity," they should comprise "structured improvisation," requiring extensive planning and development and experimentation prior the main event itself. Discussing a 2005 work by the art-research collective Sense-Lab, Manning and Massumi articulate this approach:

> The idea was that there are "techniques of relation"—devices for catalyzing and modulating interaction—and that these comprise a domain of practice in their own right. It would be the work of the event organizers to experiment with inventing techniques of relation for research-creation, not only as part of a practice of event-design, but as part of a larger "ethics of engagement." The techniques would have to be structured, in the sense of being tailored to the singularity of this event, and improvised, taking the desires and expertise of the event's particular participants into account, inviting their active collusion in determining how the event would transpire, so that in the end it would be as much their event as the organizing collective's.[93]

Eschewing conventions of audience and spectator, such events would ideally provide participants with new experiences and, more importantly, new ideas about what can be accomplished through collaboration, new "techniques of relation" that can be brought forward. The marker of collaborative success, according to Manning and Massumi, is the emergence of "a strong collective sense . . . that it was indeed possible to invent techniques for generating aesthetico-political events across a distributive network with very little central input (beyond the setting in place of a skeletal framework of enabling constraints offering affordances for cross-fertilization)."[94]

Between the mid-1970s and mid-1980s, Mobile Image developed just such a practice. Each of the three works on which this book focuses was an experiment in creating structured improvisation among collaborators—artists, technicians, organizers, users—exploring potentials for new "techniques of relation" via telecommunication technologies. The group recuperated the historical avant-garde notion of art as process, adapted for a 1970s Southern California context, in which, as Rebecca Peabody recently explained in her essay for the "Pacific Standard Time" exhibitions, artists increasingly "turned their attention from objects of art to sites of investigation."[95] The work of Mobile Image was not just collective and participatory, or what would later be called "social practice," but also a form of interdisciplinary collaborative research *as* art. Indeed, more than anything, its succession of "aesthetico-political events" comprised a distinct, decade-long research project.

The group's collaborations with technical experts and the tools of "industry" can be understood in terms of what Manning and Massumi term "immanent critique"—that is, working with (and from within) the various cultural, institutional, and economic tensions in ways that effectively "curdle 'research-creation's' annexation to the neoliberal economy." At the time, this type of ambivalent collaboration was especially prevalent in the Los Angeles area, characterized by "siteless" art networks involving distributive media, such as records, audio cassettes, radio, television, video, and print journalism, a proximity to apparatuses of mass media and communications, and the centrality of research universities that served as both incubators of experimental art and partners of aerospace and defense corporations. As Julia Bryan-Wilson explains in her essay on EZTV, the Los Angeles art scene at this time was distinguished by "alternative modes of making that [took] advantage of proximity to the resources, publicity machines, labor, and rejects of mainstream media." Bryan-Wilson describes these relationships as "parasitic," meaning that artists were both critical of industry and dependent upon it, "not strictly counter- or anti-institutional but rather oppositional in the way that so much oppositionality is really about symbiosis." Hollywood therefore "had to exist, symbolically but also materially, with its overproduction of talent and skill, to enable the growth of something like EZTV, which drafted off Hollywood workers' excess time and cast-off equipment."[96] Mobile Image relied on a similarly parasitic relationship to the emergent telecommunications industry. Galloway and Rabinowitz frequented satellite-user conferences sponsored by NASA and corporate tradeshows, where they learned how to operate advanced technologies and built relationships, recruiting experts with whom to collaborate and acquiring borrowed or donated machinery. As Youngblood explained, as part of their "research and development" practice: "Galloway and Rabinowitz work most often not with artists but with scientists, engineers, industrialists and public officials with whom they establish collaborative relationships rather than conventional vendor-client roles. They operate seriously in both the art world and the telecommunications world. They are fluent in both cultures. They can communicate with engineers, engage general systems theorists in meaningful dialog. They're semi-professional in those worlds." That said, it was also crucial for Mobile Image to maintain their position even as they took advantage of such collaborative possibilities: "The posture or stance of their interaction with industry is as important as the structures which result. The social role they play becomes an ethical issue. Most artists undertaking telecommunications projects have accepted the passive role of the client who purchases (or is donated) a pre-existing package of services that ultimately determines the structural nature of the work. In contrast, Galloway and Rabinowitz approach each project as systems integrators who work innovatively at the edge of the art and must therefore actively interface a multiplicity of tools, services and institutions to realize the nonstandard goals of their enterprise."[97]

Arguably, Mobile Image's greatest contribution to the fraught history of art and technology collaborations was their overall project's ability to go beyond the simplistic binary of collusion versus subversion noted by Bryan-Wilson, a binary that often reduces collaboration to collaborationism and thus rehearses an undifferentiated and depoliticized narrative of the historical avant-garde as a failed attempt to challenge the autonomy of art through its reintegration into everyday life. The avant-garde and its subsequent reiterations fail because art is *collapsed into* everyday life and thus subsumed under the logic of capital. Its "techno-utopian spirit," as cited throughout the accounts presented here, is easily usurped by contemporary ideologies of creative labor. Beck and Bishop describe the situation as follows:

> The fact that EAT, CAVS, and A&T have found a new significance in the twenty-first century is not because the collectivist techno-utopia promised by the Bauhaus or Black Mountain has somehow, finally, become the goal of American cultural and corporate institutions; it is because the collaborative, project-based, time-limited interdisciplinary activity advanced by those progressive organizations has been thoroughly integrated into the corporate world. Indeed, the virtues of innovation, creativity, adaptability, and collaboration are so widely promoted in the twenty-first century that they no longer refer to the capabilities of scientific of artistic elites but serve as the guiding principles of everyday social and economic life under neoliberal capital.[98]

As artist and writer Hito Steyerl argues, art increasingly occupies life: "[The avant-gardes'] hope was for art to dissolve within life, to be infused with a revolutionary jolt. What happened was rather the contrary. To push the point: life has been occupied by art, because art's initial forays back into life and daily practice gradually turned into routine incursions, and then into constant occupations."[99]

But what if the problem is not just the occupation of life by art but also the occupation of the history of the avant-garde—and its technological predilections—by its periodic revivals? Could a more nuanced account of particular avant-garde strategies provide a narrative in which the turn to the technological *politicizes* the economies of both art and industry, critically analyzing the methods and logics of production, distribution, and commodification in order to transform them? After all, endeavors such as the Bauhaus and the Constructivists becoming Productivists aimed to transform industrial production, or rely on a transformed industry, as much as they aimed to change art's function. "Collaboration," "research," "collectivism," "process," "utopia," and other terms wielded in the name of populist techno-aesthetics lend themselves to the aforementioned neoliberal appropriation when they fail to challenge fundamentally the mechanisms of material and immaterial making, when they are themselves subject to—and tools for—reification. According to Gene Ray, there is no one avant-garde; avant-gardes are plural. Yet, as mentioned, they form an "anti-capitalist vector."[100] As part of that vector, collaboration has to be more than the joint making of commodities, just as technology needs to be something other than the amplification of existing processes. As Beck and Bishop point out, the art-and-technology programs of the 1960s and 1970s were built on a fusion of the "social utopianism" of the avant-garde and US liberalism and "technophilia," a "capacious reading of science and democracy offered by liberals and by the Bauhaus, whereby science stood for a generalized stance of unbiased, egalitarian engagement with the world."[101] But "scientific thinking" and the turn toward the technological meant something very different at the Bauhaus when historicized as an avant-garde in Ray's sense rather than through the lens of postwar progressivism. A different history and future of art-and-technology emerges, one that underscores the significance of Mobile Image, both at the time and since.[102] "Collusion," "subversion," and "parasitic" denote artists' relationships to an existing techno-industry; only a narrative that depoliticizes the technological as objective and democratic can result in the inevitable takeover of "collectivist techno-utopianism" by the neoliberal apparatus.

As Melvin Kranzberg reminds us, "Technology is neither good nor bad; nor is it neutral."[103] It is oppressive when employed by capitalist institutions toward optimization, profiteering, and commodification. In the sense of the historical avant-garde, technology as a project rather than a reified work-form means to be concerned with the world in all its relationality and complexity, to analyze in order to transform that relationality and complexity—which does not mean it cannot also be poetic, pleasurable, absurd, and ambivalent.[104] As an anti-capitalist endeavor seeking to revolutionize economies of knowledge and objects, the technological is akin to "scientific thinking" as a mode of critical inquiry and practice aimed at the de-reification of the processes of production. Technology here refers to the tools and means of making as much as it does to the technological, a scientific way of thinking, related to framing, revealing, and transforming, what Sylvia Harvey terms a "looking at the 'how' and 'why' of reality."[105] To the Bauhaus, art, craft, *and* technology were to be employed in processes of "everyday-practical problem-solving," whereby Walter Gropius defined the technological as a dialectic of material and immaterial production.[106] Just as art needed to lose its false sheen of bourgeois autonomy, technological making (traced from craft labor to industry) was to be liberated from given constraints of "reproductive, innovation-inhibiting, uncreative, on the imitation-principle reliant exercise and appropriation of traditional technics" in order to forge not only new objects but also new social and productive relations.[107] Hence, "the experimental exploration of new possibilities in handling/using (*Umgang*) of materials, tools, devices, etc." was meant to consciously and actively defy given definitions and materialization of progress, function, and ownership under capital.[108] Only under the logic of the latter does Gropius's project of creating a "new human" via the "new unity of art and technology" sound like a nightmare, for it posits the modern subject as individuated user-consumer for whom progress, freedom, and democracy materialize in the reduced form of an exponentially increasing choice among commodities.[109] Similarly, the Russian avant-garde sought to challenge and expand the mechanics and mechanisms of manufacture, rather than adopt prevalent models of productivity and object utility. To make "objects as comrades," as Arvatov put it, rather than passive things to be consumed included a critical revision of the dynamics and purposes of productive processes. Progress entailed not optimized outcome but rather the historicization and contextualization of labor and bodies qua machines. Proletarian modes of making demand a new economic relationship between worker and apparatus, in both the technical and the ideological sense. In a revolutionary society, the means of production had to be comrades as well and adjust to the needs of social labor rather than usurp the latent potentials of productive tools under the traditional logic of individuation, competition, and the commodification of work for profit maximization. The work-form of the reified object (product) and of the reified tool and their application and utility is replaced by the open project of collectively changing the system of ideological and material making. The "collectivism" alluded to by the aforementioned historians of art and technology collaborations thus takes a meaning different from that of neo/liberal progressivism. As

Gregory Sholette and Blake Stimson put it: "This then is our fetish now: that the dream of collectivism realize itself as neither the strategic vision of some future ideal, of a revised modernism, nor as the mobile, culture-jamming, more-mediated-than-though counterhegemony of collectivism after modernism, but instead as Marx's self-realization of human nature constituted by taking charge of social being here and now."[110]

We argue that Galloway and Rabinowitz's collaborations with various producers and constituencies—other artists, technicians, users, social formations—were most successful when those collaborators were actively engaged in this type of historically specific "taking charge of social being." Collective social labor neither negates nor simply reproduces differences of power and experience, but rather works to make tangible the technics of social production in order to reallocate access to and ownership of its tools. This is the main reason that we primarily refer to the group as Mobile Image, defined by its expansive and ever-changing assembly of collaborators; the artists were adamant that it was never about "Galloway and Rabinowitz," and the very logic of their approach to aesthetics and technology was grounded in the critical transformation of the concept and application of the public (as opposed to private, individuated) production of meaning, experience, and selfhood. This is also why our book is decidedly not a traditional monograph. In keeping with their own method, we approach Mobile Image as a model, as a frame and lens to ponder profound questions regarding both art and technology.

Galloway and Rabinowitz coined the term "avantpreneur" to describe that method, coupling avant-garde practice to an entrepreneurial ideal. However, this was less an alignment with the concurrent development of the creative industries than the critical continuity of an alternate avant-garde history of the artist engineer in the Constructivist and Bauhaus sense. In an unpublished retrospective text co-written with Gene Youngblood in the mid-1980s, Galloway and Rabinowitz articulate this method, explicitly contrasting it to the art-and-technology programs of the late 1960s and early 1970s, which they saw as only having had access to "technologies of scale as long as they [were] producing good public relations alchemy for industry."[111] Instead, they conceptualized their work as an "aesthetic research center," with professional engineers, technicians, and operators, off-the-shelf devices, and government and corporate sponsorship, forged into a constellation of an anti-capitalist economic model that was yet to become. The avantpreneur would thus act as a "mediator between communities of consciousness participating in the design of alternative futures," allowing participants—or, more precisely, "users"—not only to take part in technological systems but also to "define the applications, define the need, define the values."[112] Within this new model of production and distribution, users cease to be consumers, for the objects at their disposal are not finite commodities but, within their historicization and the context created, use-ful only according to their use-value for and application toward social and material transformation.

Mobile Image's collaboration with technology meant engaging the latter in its given and latent potential, both as technical devices and as scientific thinking. The role of the artists was neither to aestheticize the tools of mass communication and consumption nor to rescue them from the claws of post-Fordist flexible labor. Mobile Image intended their "laboratories" to enable people to "encounter the limitations" of technology along with the opportunities it presents, effectively to politicize the utopian art-and-technology project itself. As Youngblood put it, "The politics of telecommunication has always been the undercurrent of their work."[113] This meant both modeling technology as revolutionary tool with the potential to effect a new popular politics and exposing the entrenched politics of the technological under capital. The "magic" of technology lay with the unlocked potential of existing tools to establish a fundamentally new logic of collective making. As Rabinowitz explained: "That magic was what really propelled us. But we understood that when you work with magic you have to know all the formulas. Otherwise you end up with black magic. You have to be clear about the politics. Just to say it's magic without being clear about the negative side, the politics, the reality of it, wouldn't have power. On the other hand, to reference only the cultural imperialism wouldn't be interesting either. It was a real sense of magic, which still inspires us."[114] Attempting to collapse neither art into technology nor technology into art—which would have perpetuated a binary of creative autonomy and rational progressivism in which one is ultimately subsumed by the other—Mobile Image aimed to establish a dialectic of agency and necessity, a redistribution of the ownership of the mechanisms and apparatuses that were determined by and would, in turn, inform shared present and future needs.

The work of Mobile Image belonged to a late 1970s and early 1980s moment when notions of socialism, "scientific thinking," and audience were being reconsidered. Writing in a 1982 issue of *Screen*, Harvey proposes that any serious recovery of Brecht's demand that the intellectual and creative production be not just an interpretation of the world but also an active tool of its transformation would have to take into account that times are constantly changing and, with them, art's commitment to emancipatory struggle.[115] Intellectual-artistic engagement during moments of crisis—during "dark times," to use Brecht's term—ought to be rooted in their particular complexities. According to Harvey, in the postwar period, this meant acknowledging the limited appeal and negative connotations of once-utopian concepts such as "communism" and "class struggle," in the face of their cooptation by oppressive regimes.[116] Capitalism, as many Cold War intellectuals on both the left and the right repeatedly pointed out, had, for better or worse, emerged victorious in the contest of ideologies and resources, apparatuses and technologies. In response, she calls for an updating of Brecht's "scientific attitude," which, she recalls, was intended as "a pleasurable and change-oriented form of learning." As she explains, "This attitude looks at the 'how' and the 'why' of reality, and asks how things could be different; it proposes the transformation of nature and social reality." And yet, Harvey acknowledges that, in an era when "Lindbergh's Flight" is more likely to deliver bombs than critical modern self-discovery and Galilean ruminations over the utility of progress are drowned out by calculated exploitation of resources and labor, this outlook may appear overly optimistic and teleological, to say the least.[117] Moreover, she doubts "the possibility of working for a popular left-wing culture," in which aesthetic purpose and orchestrated communal action somehow unleash the desires of those neglected and oppressed by the institutionalized tenets of bourgeois individualism. To Harvey, this seems unlikely at a moment when social movements had replaced class identity and mass culture had unified fragmented and marginalized groups as consumers rather than as agents of social change. A trajectory of manifest populisms seemed to have either replaced or made undesirable the revolutionary myth "the popular" as a source of emancipation.[118]

But times have changed once again. Since the crumbling of the Eastern Bloc in 1989, capitalist expansion has been marked by a series of ongoing crises, challenging the binary ideology of Cold War politics from both without and within. Scholars such as Alain Badiou and David Harvey have recently reconsidered the "communist hypothesis," emphasizing capitalism's growing need to legitimize its violent and increasingly visible oppressions and exploitations.[119] Digital technologies and information networks are sites of exchange and labor, at once enabling increased exploitation and alienation *and* offering potentials for the formation of new publics and constituencies, new agents in the struggle over knowledge and experience. Concerns about the role of culture, about how to wield the tools of perception, interpretation, and creativity, remain urgent. In 2009, for instance, the activist curatorial collective What, How, and For Whom/WHW reacted to the greatest recession since the Great Depression with the Brechtian inspired call to arms, "What Keeps Mankind Alive?" WHW used the Istanbul Biennial as a lens and site to pose timely, essential questions about material and immaterial needs by drawing parallels between the economic and political turmoil of the Weimar Republic and the more current potentials and limits of an increasingly interconnected world.[120] The exhibition and catalog highlighted a quote by Frederic Jameson: "Brecht would have been delighted, I like to think, at an argument, not for his greatness, or his canonicity, nor even for some new or unexpected value of posterity (let alone for his 'postmodernity'), as rather for his usefulness—and that not only for some uncertain or merely possible future, but right now, in a post–Cold War market-rhetorical situation even more anti-communist than the good old days."[121] The Biennial became a self-declared "prism" and a platform for critically engaged curatorial and artistic endeavors. WHW established relationships between perceptions of self and world, between given and potential modes of production and the utility of art and—or, more precisely, *as*—a form of new subjectivity, "where new political subjects and spheres of action are being constituted beyond representation."[122]

When, in May 2019, Facebook CEO Mark Zuckerberg proclaimed that "the future will be private,"[123] he was ostensibly talking about *privacy*, the data safety his technology would, allegedly, guarantee its users. Yet, the dystopian undertone of the message had become a reality long before. An increase in communication and information technologies has not, as promised for decades by many innovators and entrepreneurs, automatically democratized public discourse. On the contrary, the public sphere and the tools of publicity have become increasingly *privatized* in a dual sense of the word: newspapers, media, and public space are owned by corporations and organized for maximum profitability, while sites of potential discourse tend to devolve into the broadcasting of personal attitudes and opinions rather than productive critical dialogue. Social production and reproduction are not social—that is, by and for the collective(s)—but rather exclusive, customized, and hierarchical.

Sylvia Harvey's early 1980s call to revive a popular Brechtian technics of aesthetics—manifested in the concurrent work of Mobile Image, but then seemingly quashed by an overwhelming mass-consumerist culture fueled by the promise of inevitable social progress via technological advancement and the uncritical adoption of newfangled devices—may thus turn out to have been more prophetic than anyone imagined at the time. Mobile Image engaged in a contest over the political economy of tools whose numerous manifestations included the peddling of novel devices for mass consumption as well as a range of past, contemporaneous, and subsequent art—from the art-and-technology programs of the late 1960s and early 1970s to the artist network and early Internet projects of the late 1980s and 1990s, and beyond. In today's world of ubiquitous digital (re)production, global networking, and social media, the group's work continues to have profound implications for art, technology, and the politics of public and private experience. Amid rapid advances in telecommunications and entrenched myths, Mobile Image produced fantastic interventions into the direction and conception of techno-social progress. "And if you think this is utopian," to quote Brecht, "please think why is it such."[124]

1

SATELLITE ARTS: A TELEVISION OF ATTRACTIONS

In November of 1977, Mobile Image launched *Satellite Arts* (figure 1.1), a series of improvisational dances performed by individuals thousands of miles apart but assembled in the space of a composite television screen. Using the Communications Technology Satellite (CTS), a state-of-the-art US–Canadian collaboration designed to test out possibilities for broadcasting directly to individual home receivers, Mobile Image established a bidirectional link between two groups of dancers and technicians: one at the Ames Research Center in Menlo Park, California, and one at the Goddard Space Flight Center in Greenbelt, Maryland. Following a series of preliminary tests and rehearsals during the previous few months, the three-day event comprised a series of short "scores," prearranged but unscripted scenarios performed by members of the Mobilus dance troupe, an experimental collective rooted in Anna Halprin's San Francisco Dancers' Workshop. As Galloway and Rabinowitz explain in their project description, the work was "a collaboration between dance performers, technical performers and ourselves," meant to "create a base of technical information that can be applied to future, more extensive satellite performances." With two dancers in each location—Keija Kimura and Soto Hoffman on the West Coast, Nathan Stinson and Mitsuko Mitsueda on the East Coast—*Satellite Arts* used improvisation as a way to investigate critically and self-consciously the "technical problems and performance possibilities" of long-distance interaction, not only through satellite

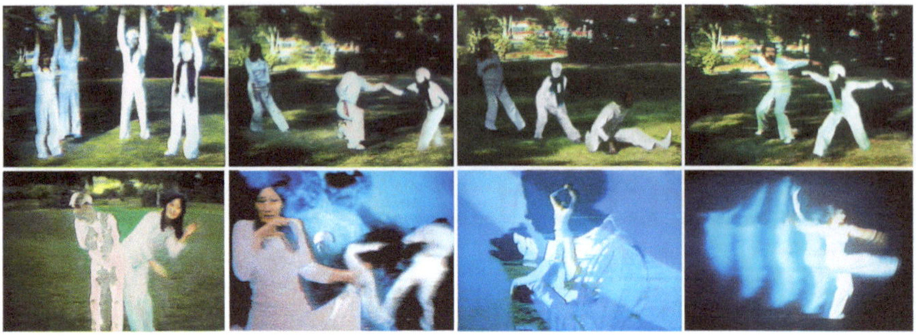

technology but also *in* "television space." Galloway and Rabinowitz clarify their approach:

When we speak of "television space," we are speaking both conceptually and literally. Literally, we are talking about a performance/dance that is designed for viewing on a television screen. In this respect, the performance is "framed" rather than staged. The camera is not an objective eye photographing a performance conceived for the stage, the dance and camera are interactive. The performance is an integration of the graphics of dance and image. Conceptually, we are speaking of a performance in which the dance/ritual is about communication, about time and space, and about our interaction with television.[1]

Satellite Arts belongs to a context of satellite experimentation, along with the broader landscape of artists engaging "TV as a creative medium," to borrow the title of a landmark 1969 exhibition at the Howard Wise Gallery.[2] Mobile Image similarly explored television's aesthetic possibilities, infiltrated its established conduits, exposed its material arrangements, and sought to open it up to bidirectional exchange. Yet, unlike

1.1

Mobile Image, *Satellite Arts (The Image as Place)*, video stills, 1977. Courtesy Rabinowitz/Galloway Archive.

many other works from this period, *Satellite Arts* expanded the communicational function of television beyond the send–receive paradigm—beyond broadcasting. Funded by a joint grant program of the National Endowment for the Arts and the Corporation for Public Broadcasting (in collaboration with NASA), intended to support experiments in interactive satellite performances, the project reimagined TV as more than a medium of symbol production and reproduction, transmission and reception, positing it as a technology of reciprocal telecommunication, as a process and a public space to be occupied and shared, through which people could meet in real time, face-to-face, even body-to-body. With *Satellite Arts*, Mobile Image tested out a pioneering counter-model that sought to fulfill the latent potential of television, which, as they understood it, "has been called a 'Window to the World' and lets us see things we might otherwise have never seen in our lives," but which was stuck in a restrictive "one-way communications system," resulting in a range of "behavioral and attitudinal habits and appropriate expectations" developed in accordance with that arrangement.[3]

The work was both a product of and a response to the rapid development of television technologies, which were simultaneously expanding capabilities and complicating the social meaning, impact, and potential of the medium. A few years prior, Raymond Williams recognized that an intensifying focus on new forms of television in the early 1970s was heralding an especially contentious and disruptive situation, namely a spike in the long-standing struggle over the media as public service, on the one hand, and as private interest and commercialization, on the other. Actual and prospective technological developments in telecommunications, including satellite transmission, cable TV, and video, were expected to shift the terms of this struggle, while likely confusing it. As Williams observed, such developments had also opened possibilities for a radical reconceptualization and redistribution of the ownership of the means of image, information, and knowledge production and distribution. Not only would conventional perspectives and narratives be challenged through a new plurality of viewpoints, but the very notion of subjectivity and selfhood in Western, capitalist society would also potentially be transformed as sites and moments, territories, and histories of experience were reflexively remediated and intertwined anew.

Engaging the former challenge while exploring the latter potential, *Satellite Arts* recalled Bertolt Brecht's appeal to transform ideological apparatuses rather than merely renovate the media that support them. Subtitled *The Image as Place*, the work provided a new site of encounter through a meaningful immediacy, a "third space" wherein performers interacted with each other, live-broadcast feeds, and the mechanical arrangement itself, generating critical connections between televisual communication and experience. Mobile Image used real-time satellite communication to establish an environment of multiple mediations *of and as* reality. The work departed from one-way distribution and from contemporaneous experiments with two-way send-and-receive models, which, while novel in their own ways, largely conformed to already-entrenched binary arrangements of participation and control. *Satellite Arts* attempted to capture the imaginative and technological potential of new technologies to set up what the artists considered "alternative social worlds as laboratories for resocialization," as places for experimentation and retraining, for repurposing mass media "as technologies of the self," as tools for finding new ways of extending the subject into the world and in relation to others.[4]

The project developed out of and departed from Galloway's and Rabinowitz's individual pre–Mobile Image work. In the early 1970s, both artists had belonged to video collectives, part of a surge of video art and activism seeking emancipatory uses of mass media's communicative tools. Such endeavors generally attempted to create new publics that could counter emergent media monopolies, whose territorial expansion was enabled by cable and satellite technologies and justified by the pervasive logic of "open skies" and "free flow" doctrines. Like a range of relatively concurrent efforts—guerilla television, "video theater," various closed-circuit and public-access ventures—*Satellite Arts* re-functioned available sound and imaging devices to create an intimacy and immediacy of perception, both stoking and interrogating what Walter Benjamin noted as "the desire of the masses to 'get closer' to things spatially and humanly."[5] Yet, the work stood in contrast to efforts that sought to expand, and in the end reproduce, the broadcast logic of the existing media apparatus, however well-intentioned in their goals to "democratize" or "humanize" a presumably monolithic consciousness industry.

At a moment when television was rapidly extending its panoramic reach via new technological developments and promises of expanded participation and global connection, *Satellite Arts* engaged the televisual as "symbolic form," as a dominant representational scheme that both liberates experience and imprisons the subject within a preset perceptual regime. The work aimed to make palpable the particular organization of connections and disconnections between people, spaces, and moments under that televisual regime in order to highlight the latent and dormant potential of communication technologies and to enable their emancipatory use. Rather than simply convert receivers into senders, consumers into producers, the work allowed participants to experience televisual time and place as one of configured, and therefore reconfigurable, relationality. The goal was not some fanciful, communalistic zone of unalloyed connectedness but rather a laboratory in which participants could playfully test out—and try out—both the potentials and the limitations of a technological reconfiguration of spatial and temporal relationships, of always-already mediated selves and others. This was a new public space that was self-consciously hybrid or "composite," in Miriam Hansen's use of the term, that is, "an effect both of the material—economic, technical—imbrication of distinct types of publicity and, concomitantly, of the coexistence of multiple, interacting and competing horizons of experience on the level of individual consciousness."[6] Mobile Image at once activated a materialist critique of the medium and its ideologically bound divisions—public–private, active–passive, sender–receiver, producer–consumer, real–virtual—and modeled forms of individual and social subjectivity, conceived not as stable, coherent wholes but rather as interwoven amalgams of multiple, shifting, at times even contradictory, perspectives and relations.

Satellite Arts enacted a consciously theatrical performance of interacting bodies, of multiplied selves in layered space, communicating in and through the screen, via delayed sounds and images—an assemblage of limbs, sights, and attitudes, at once organized and improvised. The work's reflexive, participatory approach to the medium, its politics of perception, recalled what Tom Gunning termed the "cinema of attractions" of the early twentieth century, an immersive, fairground-like approach to filmmaking and display that theatricalized its own modes of visuality, prompting relational negotiations of mediated meanings rather than absorption into a single self-contained and illusory cinematic world. In retrospect, this paradigm of mobilized viewing offers an alternative to the subsequent establishment—and ultimate entrenchment—of fixed forms of noncommunicative, disembodied spectatorship, physically and socially isolating, simultaneously massifying and privatized, as exemplified by notions of TV as an industrially produced window onto the world. Such entrenched spectatorship facilitates the extension of televisual logic both locally and globally, enabling its neo-imperial reach by enshrining territorial distinctions between "here" and "there" and reinforcing control over information and knowledge production. A kind of "TV of attractions," *Satellite Arts* empowered participants to consider the broader televisual apparatus, experiment with the rearrangement of its structures, and inhabit a specific site of shared physical, intellectual, and social experience that was itself set in relation to other spaces of experience. A trenchant precursor to contemporary forms of network politics and aesthetics, it provided a particular organization of mediation, of technological relationality as inter- and intra-subjective process. Although certainly limited in scope and practical feasibility, the work fostered a self-consciously politicized, potentially transformative engagement with satellite technology, not only as a means to an end but also as a tool of both connection and separation that can be repurposed to unlock new possibilities of communication and assembly, of public and private formation.

THE TELEVISUAL REGIME

During the late 1960s and 1970s, artists, activists, and theorists were enthusiastic about new developments in televisual communications technology. The proliferation of stations and in-home receivers, of satellite transmission, cable distribution, and video equipment, led Gene Youngblood to declare TV a "sleeping giant," ready to be used "in new revolutionary ways," and Raymond Williams to proclaim that "a world-wide television service, with genuinely open skies, would be an enormous gain to the peoples of the world."[7] At the same time, writers and experimental practitioners were grappling with television as a complex and often contradictory form of politics—as a technological organization of experience determined by what at the time was perceived as an ambiguous yet hegemonic televisual regime. This regime is maybe best described by Hans Magnus Enzensberger's analytic extension of the "general contradiction between productive forces and productive relationships" onto "electronic media," hence between material and immaterial means of production and the relationship of producers—or, in this case, users—to those means of production and to one another.[8]

Although much art of this period can be understood in relation to television both as a mass medium and as an increasingly dominant cultural influence, only a relatively small cohort of artists actually used it in their work. Such work pointedly altered normative television arrangements, from its prevailing aesthetics to standards of representation, distribution, and viewership. Nam June Paik's early engagements with television involved various attempts to treat the TV as an object—one that occupies physical space and can be interacted with, altered by external forces, or even affixed to the body. Beginning in the early 1960s, Paik attempted to activate reception, upending visitors' typically passive relationship with the TV screen.[9] In the late 1960s, public television studios such as KQED in San Francisco, WGBH in Boston, and WNET in New York established local "TV labs," where artists-in-residence were encouraged to experiment with television production and rethink its purpose and function. Although diverse, the results mainly consisted of works that either treated the screen as a painting-like surface—kaleidoscopic compositions involving superimposition, visual echoing, and other manipulations of color, space, sound, and form—or explored TV's multimedia potential, melding dance, music, or poetry with televisual effects.[10]

Other artists built upon Paik's early work, attempting to "materialize" the television apparatus and its processes of mediation more emphatically. They moved away from formal experimentation and instead drew attention to the mechanisms of TV operation—both machinery and audience. Frank Gillette and Ira Schneider's *Wipe Cycle* (1969) was a bank of nine monitors with a hidden camera pointed at the viewers themselves (figure 1.2). The system captured a live feed of those standing before it, and then cycled that imagery, with various levels of delay, through the grid of monitors, mixing the feed with recorded segments of

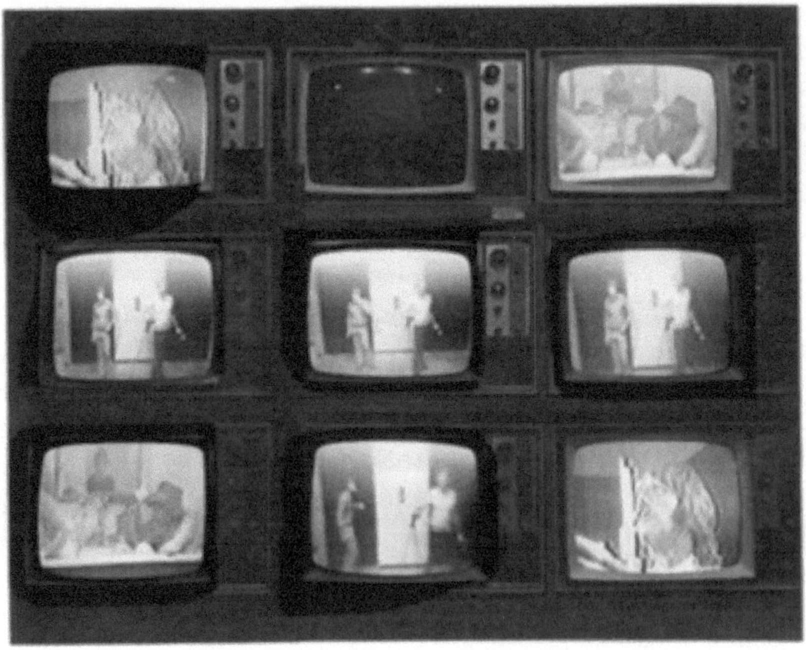

broadcast television and diverse video footage compiled by the artists—views of the Earth and outer space, urban and rural scenes, a "skin flick" bathtub scene.[11] Spectators thus watched themselves not only in the act of watching but also instantly transformed into one reproduction among many within a televisual universe. By positioning the audience in a self-conscious relationship to the broader media ecosystem, the artists sought to "demonstrate that you're as much a piece of information as tomorrow morning's news," as Gillette explained, "in other words, rearranging one's experience of information reception."[12] In 1976, Michael Asher spotlighted the material conditions of television production as well as reception. His contribution to the Portland Center for the Visual Arts'

1.2

Frank Gillette and Ira Schneider, *Wipe Cycle*, 1969. Courtesy Frank Gillette and Ira Schneider.

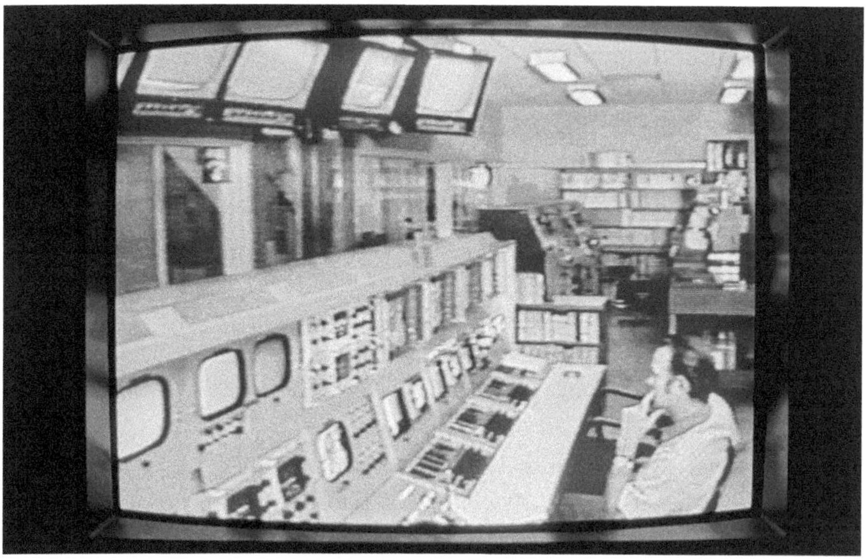

exhibition *Via Los Angeles* was a thirty-minute broadcast of the master control room of the local KGW station (an NBC affiliate in Portland), on which that footage was aired (figure 1.3). As with his other works from this period, Asher focused on the usually hidden physical components of the institution—in this case, tape decks, switching panels, technicians, and banks of monitors, one of which contained the camera's own image, while others showed recorded programming set to be televised later that day. As Asher put it, "Viewing these images, the audience realizes the degree of mediation necessary to the production and reception of TV images. The audience also understands that the TV image is an electronically generated depiction of real space on a flattened plane at a reduced scale with

1.3

Michael Asher, Portland Center of Visual Arts, Portland, Oregon, USA, *Via Los Angeles*, January 8–February 8, 1976, documentation of the live feed of the control room at KGW (radio and television). Photograph by Michael Asher. © Michael Asher Foundation.

light and sound representations recorded by camera and sound equipment." Asher additionally intended this reflexive project as a response to the ascendance of video art at the time, an attempt to shift attention to "the mode of production from which it [video technology] originated: television technology."[13]

Such efforts to examine the operations of televisual mediation belonged to a moment in which a core contradiction was becoming increasingly clear: alongside rapidly advancing technical means that allowed for greater immediacy, range, and choice of programming, entrenched televisual norms perpetuated depoliticized, disembodied, and isolated forms of spectatorship. This contradiction was sustained by the broadcast paradigm, what media activist Brice Howard called TV's "logic of distribution." Principally concerned with the power of information dissemination, this logic effectively crippled the *communicative* potential of the media, focusing instead on access to the channels of unidirectional delivery.[14] The challenge therefore was to extend a critical discourse and practice of the televisual beyond notions of the medium as a technical and material machine to an understanding of the televisual as a dominating and potentially emancipatory apparatus of sensual, sociopolitical, and economic operations.

Certain artistic interventions struggled with this challenge, attempting to reconfigure conventional modes of television production and reception actively, including TV-based work that incorporated audience participation. Presented as the "world's first participative telecast," Douglas Davis's *Electronic Hokkadim* (1971) occupied thirty minutes of broadcast time on a Washington, DC, television station. When viewers called into the station, the sounds they made were routed through various devices that translated them into glitches on the screen.[15] Other artists attempted to "democratize" the technology by expanding access to broadcasting rather than enabling audience participation. "Guerilla television" collectives such as the Raindance Corporation, Videofreex, and Top Value Television (TVTV), for example, embraced decentralized means of distribution made possible by the emergence of cable, along with do-it-yourself production made possible by increasingly obtainable and portable video equipment. These groups pursued more "objective" content, explicitly in opposition to what was seen as the overly slick, aesthetically bankrupt, and corrupt world of commercial media.[16]

Similar attempts to reallocate the means of television production characterized the few mid-1970s projects that, along with *Satellite Arts*, constituted the first satellite-based works of art. In 1976, Davis broadcasted the final part of his performance *Seven Thoughts* from the Houston Astrodome via a commercial cable television satellite. Standing in the empty stadium, he sent out "seven very personal thoughts to the citizens of the world." Davis saw his use of satellite as a proclamation "that long-distance broadcasting was not an exclusive preserve of TV networks, armies, and governments,"[17] while the intimacy of his transmissions was meant to contrast with "imperialistic" media content.[18] The following year, as part of documenta 6, Davis produced *The Last Nine Minutes* (figure 1.4), transmitted live to more than twenty-five countries, along with performances by Joseph Beuys, Nam June Paik, and Charlotte Moorman. Although the very idea of satellite TV as a forum for art was groundbreaking in and of itself, only Davis engaged the medium directly. (Paik and Moorman showed some of their famous earlier works; Beuys lectured on his concept of "social sculpture.") The performance opened with what appeared to be a viewer's hands banging on the surface of his home television to get the attention of the artist, who is seen in the middle ground. Davis then approached the viewer and pounded against the screen himself, with increasing agitation, for the remainder of the piece. The work underscored the spatial separation between audience and performer, in contrast to their shared temporality, which was emphasized by the soundtrack of a ticking clock and the stopwatch hanging around Davis's neck at the start of the performance. As Kris Paulsen explains, "Davis dramatized the desire to use the technology to 'reach out and touch' the viewer, but the hands that seemed to appear on the viewer's side of the screen only served as an uncanny reminder of the media's limitations."[19] Also in 1977, Liza Béar, Keith Sonnier, and a host of collaborators produced *Send/Receive* (figure 1.5), designed as a hybrid of live video conference and improvisational performance. The first part of this project was a twenty-three-minute prerecorded video broadcast on a New York cable television station, featuring a cacophonous mix of materials—photographs, audio clips, maps, charts, historical documents, and scrolling informational text—intended as both a dialogue in support of public access to satellite technology and an example of the technology's potential for information

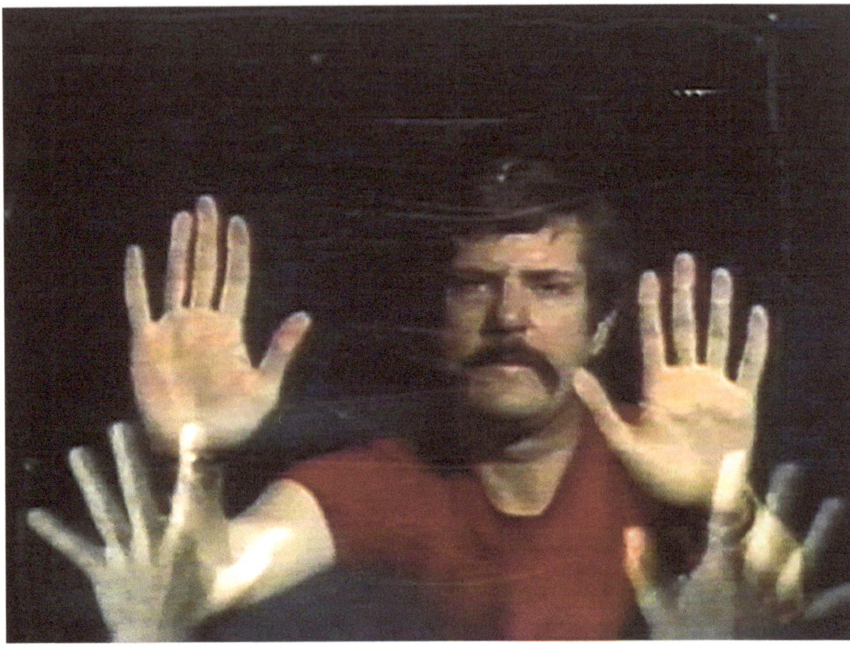

overload.[20] For the second part, artists in New York and San Francisco participated in a live bidirectional satellite transmission, whose results were edited together in real time and broadcast on cable television stations in both cities on a split screen (in which two views are shown simultaneously on a vertically divided screen), one side for New York, the other for San Francisco. Technical difficulties plagued the work, and planned collaborative performances were replaced by descriptions of how the technology functioned, arguments for public access, and overlaid text explaining the technical problems at hand.[21] *Send/Receive* explored the potential of (and, as it turned out, the impediments to) bidirectional television, while making a polemical case against centralized control of broadcasting. To

1.4

Douglas Davis, *The Last Nine Minutes*, still of performance for international satellite telecast at documenta 6, 1977. Courtesy Electronic Arts Intermix (EAI), New York.

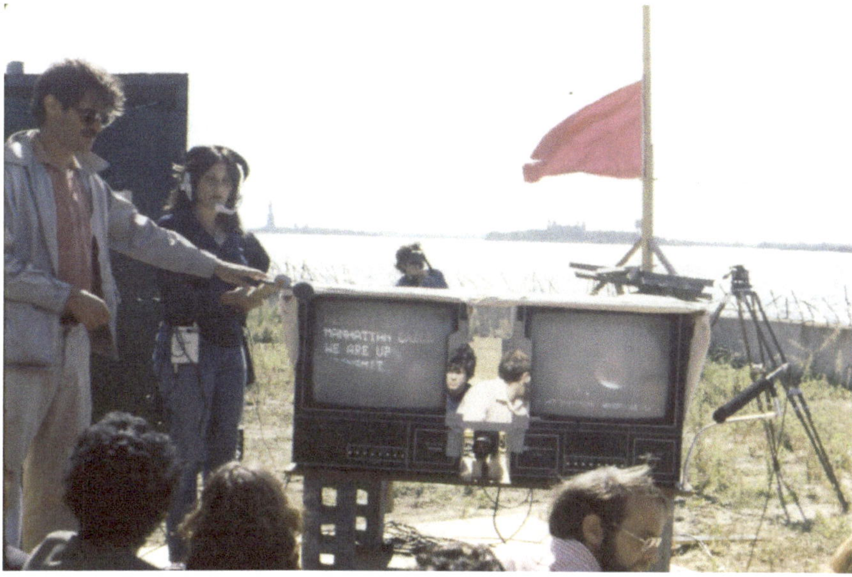

the artists, the work's content was less important than the very act of accessing and using satellite technology. As Sonnier explains, "Acquiring that tool was the political thrust—making that tool culturally possible."[22] That said, as with *Satellite Arts*, *Send/Receive* did not explicitly address the broader politics of access and its uneven manifestations—who had access to such tools (and requisite funding sources) and who did not.

The contest over not only technical devices but also perceptions of communications technology as a constitutive part of a specific social structure had thus evolved since the 1960s, when, according to Enzensberger's influential 1970 essay "Constituents of a Theory of the Media," the New Left had painted too simplistic a picture of an oppressive mass media apparatus in general and TV's role as part of it in particular. Dismissing the Orwellian "bogey of a monolithic consciousness industry" as "undialectical and obsolete," Enzensberger instead calls for an understanding

1.5

Liza Béar and Keith Sonnier, *Send/Receive*, 1977.

of communication, knowledge, and experience as always "manipulated" and thus the abandonment of depoliticized quests for "pure" interaction and knowledge in favor of a critical theory of media utility and relationality.[23] This demand for an analysis of perceptual connectivities established by and through the televisual is echoed by Allison Simmons in her introduction to *The New Television: A Public/Private Art* (1977), in which she describes TV as having "diffused our understanding of past and present," "fostered an ambivalence between activity and passivity," "blur[red] the distinction between the real and the fictitious," and "intertwined the traditional concepts of public and private."[24] A critical media theory and practice would have to come to terms with the televisual beyond the technocratic-versus-"late liberal" binary, as something more complex than either an inescapable totalitarian imposition of false consciousness or a collection of neutral devices to be reappropriated and wielded at will.

When discussing the question of whether or how new technological developments would yield "alternate uses," Williams assesses the situation as follows:

> There can be little doubt that in the early 1970s we are already in a new generation of communications technology, and that much of this is centered on new forms of television. At the same time we are in a very contentious and confused situation about the institutions and social processes of all communications. There is still an unfinished struggle and argument over the institutions and control of sound and visual broadcasting; the conflict that has been clear for two generations between "public service" and "commercial" institutions and policies. It would be a major error to suppose that this conflict is over; indeed the signs are that it is now entering one of its most acute and difficult phases. But at the same time the actual and prospective development of new kinds of technology is altering some of the terms of this long-standing conflict, and may, if we are not careful, merely confuse it. *On the other hand, some of the new technical developments seem to open the way to institutions of a radically different kind from either "public service" or "commercial" broadcasting; indeed of a different kind, in some cases, from "broadcasting" itself.*[25]

The conundrum was how to use new technologies—cable systems, satellite communications, portable recording devices—in a political, critically reflexive, and thus truly transformative manner rather than merely perpetuating existing ideological institutions and their ongoing materialization of utopian as well as dystopian techno-myths of unbridled liberation, on the one hand, and total domination, on the other. What would it mean for technical developments to initiate practices "radically different," as Williams put it, "from 'broadcasting' itself"? Technology as a social tool and social construct would then have the potential to unsettle traditional categories of public and private, of ownership of the means to produce and organize experience through information and sensibilities that establish and imbue with meaning relationships between self, other, and world. To go "beyond broadcasting" would be to construct a media practice based on a politics of perception, a dialectics of media as *communicative*, beyond the options to either send or receive, toward not only questions of access, participation, and self-expression or consumption, isolation, and control but also an understanding of these things as thoroughly interrelated mechanisms. When Oskar Negt wrote in 1973 that "the media do not constitute the core of a critical media theory," he emphasized a *technics* of communication over its technical means, of communication instruments as always actively inscribed with history, meaning, and function (rather than perceived as somehow beyond ideology, as somehow neutral or possessing inherent qualities—liberational, oppressive, or otherwise).[26] Public and private consequently become relative, inter-determined, continuously negotiated categories instead of traditionally and ostensibly autonomous spheres artificially compromised by media manipulation. To Negt, communication is "publicity"—a social process and struggle over meaning and identity, territory and history determined first and foremost by the relationship between private and public, between individual and social subjectivities.

Mobile Image's refunctioning of television into a medium of reciprocal telecommunication situates it within a broader framework that includes technologies such as telephone and radio, their early histories as well as their then-current conceptualizations and uses. In *Understanding Media: The Extensions of Man* (1964), Marshall McLuhan describes the telephone a means of "extra sensory perception" that was initially considered a potential mass medium, a "PA system" that could rival newspapers.[27] In fact, from its beginnings in the late nineteenth century through the mid-twentieth century, telephone was primarily conceived as a communal experience, lauded for its unifying and democratizing potential. "Party lines" were open to anyone, providing access to a range of broadcast content, including religious sermons, news, and music.[28] This arrangement was eventually abandoned, and telephone rapidly became almost exclusively a system of private, one-to-one conversation, enabling people effectively to be in two places at the same time but at the expense of collective experience. It ultimately became, as McLuhan noted, "the most removed of any medium from the PA form."[29]

Accordingly, for the few artists who have subsequently taken up the technology, it has served as a tool of information transmission across spatial divides, assisting conceptualist efforts to remove the hand of the artist and distribute art-production processes. "Art by Telephone," for example, was a 1969 exhibition at the Museum of Contemporary Art, Chicago, loosely based on Laszlo Moholy-Nagy's 1923 "telephone pictures," a series of paintings whose compositions the artist claimed to have dictated over the phone. For the MCA show, artists from the United States and Europe were asked to call in their ideas for temporary artworks, which would then be executed by museum staff, with the artists' oral instructions exhibited alongside the finished work.[30] Here, the technology was used unidirectionally—messages sent and received—like an elaborate game of telephone. That same year, for "When Attitude Becomes Form" at the Kunsthalle in Bern, Switzerland, Walter De Maria exhibited *Art by Telephone*, which held the prospect of a private conversation between artist and viewer. The work consisted of a black telephone placed on the floor beside a sign that read: "If this telephone rings, you may answer it. Walter De Maria is on the line and would like to talk to you."

Like the telephone, radio evolved from a multipurpose telecommunications technology to a mainly singular one. It initially was devised and used as a tool of bidirectional, one-to-one transmission between amateur operators—a form of "wireless telegraphy," as it was called—and it occasionally even served as a public forum in which many voices could speak. To the historical avant-garde and the proletarian movements of the early twentieth century, the radio was an essential device for the formation of critical, emancipatory consciousness. Writers such as Brecht, Benjamin, and Alfred Döblin wrote about and for the medium, while worker's organizations in Germany, the Soviet Union, and elsewhere established radio clubs with the goals of exerting influence over existing programming and establishing proletarian broadcasting stations.[31] Brecht's famous and much-quoted essays on the radio emphasized the potential of the technology, differentiating specifically between the mechanisms of broadcasting, on the one hand, and two-way communication, on the other. Arguing that there was no inherent function to the device and that it thus could be reappropriated by inverting its reigning logic, Brecht (re)defined the meaning of communication from the distribution of information to the social production of knowledge. He famously lambasted the celebration of expansionist attitudes, of the mass distribution of irrelevant subject matter under the guise of public culture—"the colossal triumph of technology to finally make Viennese waltzes and cooking recipes available to the entire world"—as leading to the isolation of the listener.[32] Instead, according to Brecht, "The radio would be the finest possible communication apparatus in public life, a vast network of pipes. That is to say, it would be if it knew how to receive as well as to transmit, how to let the listener speak as well as hear, how to *bring him into a relationship instead of isolating him*. On this principle the radio should step out of the supply business and *organize its listeners as suppliers*."[33] As technological advances were propagating new models of social formation and interaction, Brecht urged relationality, in a scientific-pragmatic as well as intellectual sense: he advocated the ability literally to talk back, to speak, to discuss via the apparatus of transmission, yet also implied that *what* is being communicated must have relevance to the listener, must implicate and engage him or her as a speaker not to but *of* the public.

Nonetheless, aside from limited exceptions, the radio has been standardized into a medium of predominantly one-way, one-to-many broadcasting.[34] Arguably, its more overtly public role as a technology of mass distribution explains why artistic engagements with radio, while hardly plentiful, have been more common than those involving the telephone. Several artists have made use of the abundant transmissions traveling the airwaves, highlighting both the ubiquity of radio and its local specificity. In John Cage's *Imaginary Landscape #4* (1951), for example, twelve radios are "played" by performers—two on each, one controlling the station dial, one controlling volume and tone—according to the composer's score. Cage was drawn to the radio as a musical instrument because its sounds were arbitrary, dependent on the location and choice of whomever was tuning in at a particular time. Jean Tinguely's early 1960s "radio sculptures" comprise various found materials, motors, and dismantled but still functioning radio parts made to switch continuously between stations, playing bits of local broadcasts. As in Paik's television works from the same period, Tinguely's sculptures emphasize the technology's materiality. And, as in Cage's composition, the radio serves as an index of a specific place and time, while incorporating public transmissions into otherwise autonomous works of art. Similar effects were produced by Keith Sonnier's *Scanners* (1975), which consists of six radio scanners tuned to various frequencies so that they receive whatever real-time transmissions happen to be passing through the room at the time—from police dispatches to weather reports to ship-to-shore messages. *Scanners* divulges the plethora of invisible signals that permeate modern life.[35]

Others during this period attempted to repurpose radio technology and open up its participatory and communicative potential. Max Neuhaus's *Public Supply I* (1966) brought together the telephone and the radio so that the public could participate in a live sonic performance. At a time that predated call-in talk shows, listeners of WBAI in New York were encouraged to call into the station, where the artist mixed whatever sounds they made and broadcast the results out over the airwaves. "I realized I could open a large door into the radio studio with the telephone," explains Neuhaus. "If I installed telephone lines in the studio, anybody could sonically walk in from any telephone."[36] The artist also played with the feedback produced by listeners whose radios were on when they called in, materializing the relationship between signal and receiver, broadcast and audience. A decade later, Neuhaus produced the more ambitious *Radio Net* (1977). Using the broadcasting loop connecting 190 National Public Radio stations across the country, he orchestrated a collaborative symphony produced by the work's listeners, brought together in what the artist termed a "virtual place."[37] Neuhaus opened up broadcasting to audience participation and to collective sensory encounter, enabled by the hybridization of two century-old telecommunications technologies. As Dieter Daniels put it, the work "restag[ed] the 'relatedness' between radio and telephone, lost in the mists of media history."[38]

Other attempts to repurpose radio technology involved the establishment of regular programming—or at times entire radio stations—either designated for the distribution of experimental art or conceived as works of art themselves. In 1970, Berkeley's KPFA began broadcasting the "World Ear Project," a program of ambient recordings made by listeners, "an attempt to bring all our ears closer together."[39] Everyday environmental sound was the sole content of the series. "Close Radio" (1976–1979) on KPFK in Los Angeles was a weekly half-hour radio show that allowed artists to present works of sound art. Billed as an "audio space for visual artists," its founders saw radio as a largely untapped public arena for performance, a way to deliver art to mass audiences by circumventing exclusive art-world institutions (art magazines, galleries, museums).[40] Works included abstract sound compositions, noise music, sound collages, spoken-word pieces, lectures, and extemporaneous interactions with the public.

Seen in relation to this broader context of telecommunications practices, *Satellite Arts* can be understood as having combined possibilities of broadcast television with those latent in older technologies such as telephone and radio—operations that had become largely obsolete but which were occasionally excavated in select art projects or could be glimpsed in more specialized uses, such as the conference call, two-way radio (walkie-talkies, CBs, dispatchers, etc.), and call-in talk shows. In this sense, the work was as much a recuperation and merging of overlooked potentialities as it was an embrace of new technological capabilities. Galloway and Rabinowitz recognized this at the time, writing in their notes for the improvisational performance project "maintained analogies of 'Radio' and 'Telephone,'" specifically with regards to "the spontaneous-interaction that distinguishes 2-way Communication from One-way-Broadcast."[41]

Yet, *Satellite Arts* was not just about transforming television from uni- to bidirectional, from a medium of reception to one of audience participation, allowing consumers to take part in production. Whereas Galloway and Rabinowitz certainly understood their work in relation to other telecommunications technologies, they were also wary of seeing new possibilities as mere extensions of older ones. Regarding contemporaneous attempts at participatory media, the artists complained that "if television has been called 'radio with pictures' then conventional applications of interactive tele-communications can be called 'telephone with pictures.'" To them, the establishment of a composite televisual space in which people could come together broke with then-current practices, not only standard one-way transmission but also emergent experimental approaches, which they saw as largely limited to two-way information exchange.[42] *Satellite Arts*' expanded notion of communication—beyond the broadcast—more fully satisfied the redefinition of the term that initially accompanied the rise of electronic media in the mid-nineteenth century. As John Durham Peters explains, the emergence of such media at that time "refitted the old term 'communication,' once used for any kind of physical transfer or transmission, into a new kind of quasi-physical connection across the obstacles of time and space."[43]

As both a cultural phenomenon and a set of socioeconomic operations, the broadcast logic is rooted in bourgeois society's construction of subjectivity through publicity and thus encompasses, historically and technologically, much more than television alone. In 1980, the historian Stephan Oettermann argued that "the television of today is a direct descendant of the panorama," the nineteenth-century immersive circular spectacle of painted land- and cityscapes. Oettermann's account of what he deems the "first true visual 'mass medium'" provides an insightful lens onto the complexities of the various 1970s efforts to engage the televisual regime critically.[44] This account is essentially the history of a way of seeing and positioning that found an embodiment in, and perpetuation through, a powerful cultural device that allowed its audience to embark on adventurous journeys and occupy dizzying vantage points without ever having to enter a foreign land, scale a mountain, or climb aboard a hot-air balloon. Combining the modernizing powers of past and present innovations such as photographic representation, railroad transportation, and ocularcentric perspective, the panorama let individual and collective bodies extend beyond physical limitations of sensibility, instilling in a growing, aspiring Western and urban populace a new perceptual competence and dependency—both exploitative through its penetrating, invasive reach and "imploitative" through the subject's permeation with mass-produced and highly orchestrated perceptions. As part of the enlightenment's promise of individual emancipation and social progress, the panorama enabled mass participation in the discovery and occupation of new territories and emboldening perspectives, appealing to popular desires of independence, entertainment, and mobility.

Crucially, however, "televisual" technologies such as the panorama (and, by extension, TV itself) permit some forms of movement over others. According to Oettermann, the panorama, as "an apparatus for teaching and glorifying the bourgeois view of the world," literally and symbolically expanded and contained its audience's horizon, serving "both as an instrument for liberating human vision and for limiting and 'imprisoning' it anew."[45] This bourgeois view helped to contain and control that desire of the masses, as Benjamin noted, to "get closer" to things.[46] The televisual isolates the viewer and dematerializes the gaze, producing a subject that, although ostensibly part of a spectating crowd, finds its individuality affirmed while partaking in a democratic quantification of knowledge through consumption. Technology is the tool that expands the horizon while keeping it at arm's length, establishing a simultaneously remote, controlled, and unidirectionally intimate observer's relation to other people and places, mobilizing the disembodied, roving eye while centering and anchoring the subject as an emphatically private (rather than consciously social) and dissociated owner of experience. Jonathan Crary describes the unifying and fragmenting dialectic of the panoramic viewpoint: while providing "an imaginary unity and coherence to an external world that, in the context of urbanization, was increasingly incoherent . . . , the panorama was in another sense a derealization and devaluation of the individual's viewpoint."[47] Culminating in television itself, the televisual regime is both "mastery of a position that transcended local provincial viewpoints" and an expression of "the tragic insufficiency of the relation between the subject and the world," a manifestation of "new forms of subjective isolation, of a sensory impoverishment and emotional privatization."[48] As Caroline Jones has pointed out, this came to a head in the late 1960s and 1970s, when the "technological estrangement of the senses" was increasingly identified beneath a range of problems, from oil crises to processed food to televised war.[49]

Herbert Schiller saw the rapid development and dissemination of communications technology under the televisual regime as a form of intra- and international imperialism. In *Communication and Cultural Domination* (1976), Schiller emphasizes the relation between technical means and a politics of technology: "Technology and the way it is used affect the basic infrastructure of social communication. Thus the acceptance of a 'developmentalist strategy' in a nation introduces more than just industrial techniques and equipment. The way human beings are related to each other in their work and their community and family life is largely, if not overwhelmingly, determined by the nature of the technology employed, how it is employed, and the social relations that govern its use."[50] The "free flow" doctrine promoted by the United States after World War II entailed a material and ideological strategy of economic expansion. "Freedom of information and movement were highly desirable and legitimate aspirations of occupied nations and peoples. And it was relatively easy to confuse truly national needs with private business objectives."[51] According to Schiller, the globalization of mass media expanded the panorama as well as the panoramic. Presumably, nonideological devices and machines promised to create a more inclusive, more democratic global public sphere while extending the reach of particular discourses and products in the search of new markets; "free flow" translated de facto into a unidirectional delivery system of material and immaterial goods. "A largely one-directional flow of information from center to periphery represents the reality of power," Schiller explains.[52] These forceful developments included both the creation of new consumers and markets and the exportation of attitudes, perceptions, and outlooks: "Assisted by the sophisticated communications technology developed in the militarily oriented space program, techniques of persuasion, manipulation, and cultural penetration are becoming steadily more important, and deliberate, in the exercise of American power . . . Made-in-American messages, imagery, life-styles, and information techniques are being internationally circulated and, equally important, globally imitated."[53]

The "free flow" doctrine even informed efforts meant to treat the disparities and disproportionalities arguably caused by it. Indeed, the stated focus of NASA's "Public Service Communications Satellite Experiment," the program that provided access to the technology for both *Satellite Arts* and *Send/Receive*, was to extend the reach of the satellite and its broadcast possibilities (figure 1.6). In an attempt to counter the overcommercialization of the technology, this program was "devoted to experiments and demonstrations of various public services" via the application of satellite technology for "the betterment of life on Earth." "Betterment" here was

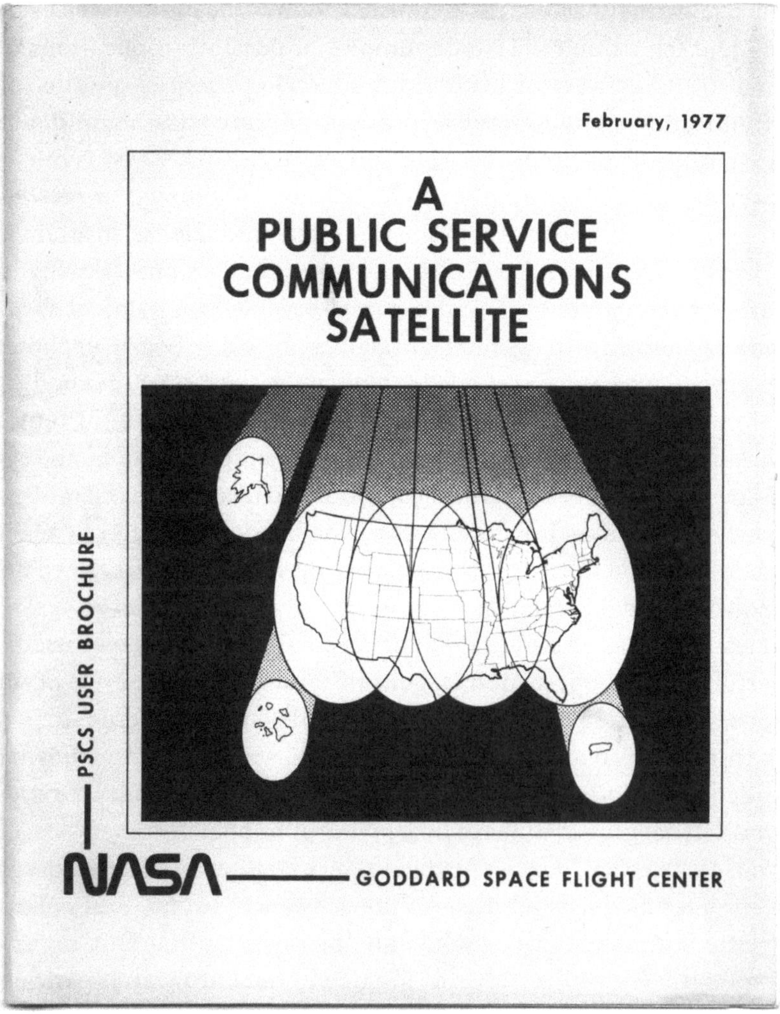

1.6

NASA, "A Public Service Communications Satellite," February, 1977. Courtesy Rabinowitz/Galloway Archive.

explicitly understood in terms of the satellite's ability to reach "remote regions . . . without access to the normal communications services (telephone and television)."[54] Listed examples of possible applications were limited to the delivery of health and education services, public safety communications, and expanded information transmission, and all but the two art projects proposed were conceived in such ways. While they took advantage of the program, Galloway and Rabinowitz were decidedly opposed to its underlying premise, understanding it as an attempt by NASA to "justify their existence" in the face of budget cuts. Having lived in Europe from 1969 to 1976, Galloway also saw his own political awakening as a reckoning with US media practices abroad—the use of "Voice of America" broadcasts to bolster US foreign policy and, more generally, the dissemination of American mass media products as a form of "cultural imperialism"—all of which he saw as "American propaganda about the 'free flow of information.'" This recognition led to a split in the mid-1970s with his then-collaborator, Jack Moore, who had joined a UNESCO effort to "declare war on illiteracy" by transmitting video tape directly to undeveloped countries via satellite.[55]

Lisa Parks has shown in great detail how the satellite was used as a powerful and strategic symbolic, cultural, and economic device of transnational unification and exploitation—and Paulsen references Park's work in relation to artists using satellite TV, specifically *Satellite Arts*.[56] So-called satellite spectaculars, 1960s live-via-satellite international TV programs, provide examples of how myths of "global presence" served to engrain the televisual regime further.[57] The 1967 intercontinental program *Our World* reached five hundred million viewers in twenty-four countries in an attempt to demonstrate the unifying powers of satellite broadcast technologies. The project was coordinated by the European Broadcasting Union and master-edited by the BBC London, and although it reached as far south as Australia and as far east as Japan, *Our World* effectively excluded developing nations as both producers and receivers, while the Soviet Union and Eastern Bloc countries withdrew their participation at the last minute as a form of political protest. The presence of those without the resources or power to participate was reduced to graphs, statistics, and prerecorded images.

Although the program took years to plan and was meticulously rehearsed, the producers of *Our World* theatrically foregrounded "liveness" and "immediacy," complete with playful delays in transmission and celebratory visualization of the technical complexity of such unprecedented "progress," to stage dramas of humanist proximity that told stories of a "Great Family of Man," while never ceasing to remind viewers of their place in the world. The "live" birth of four babies in different places—Japan, Mexico, Canada, and Denmark—allowed viewers to see themselves as part of a global community rooted in the most basic and indisputable commonalities; the presentation of the child born prematurely in Mexico City and the behind-the-scenes demonstration of the Mexican TV crews struggling to comply with the technical capacity to partake in this collective implied a notion of difference based on potential and choice, on the apparent sheer willingness to be a receiver of the civilizing forces and attitudes necessary for well-being and success. As Parks explains, "*Our World* produced a global mapping of technological development, dividing the world into zones of technological progress and illiteracy, and used the liveness of the satellite-relayed signal to dramatize and reinforce them."[58] For a country or place to partake in such real-time transmission meant to be modern, to be civilized. Meanwhile, the windows opening onto and from such sites did not necessarily offer new forms of perception but rather a leveling expansion of existing Western viewpoints of modernization onto community and humanity, difference and otherness. While geographical limits of sight toppled under an ever more omnipotent God-like view, the social horizon of experience largely remained the same, was strengthened even; the televisual gaze affirmed binaries of center/periphery, north/south, developed/traditional, self/other, public/private. As a form of perceptual tourism, the newly mediated reach promoted a disembodied, distanced, and distancing form of mobility that was equated with freedom and liberty, while being connected and visible, "live" and simultaneous, became the hallmark of integration into the so-called developed community of nations, the symbolic "we" of a planet as humanist habitat, visible from outer space—images of unity provided by the latest and most advanced broadcast technology. Parks asserts the new media's panoramic dynamic, linking the bodily and spatial rhetoric of techno-progress with its public, political function, as was espoused by enthusiastic European and US politicians and intellectuals at the time.[59]

The power of the televisual and its broadcast logic thus has at its core a relationality that divides and distances the public and the private through a familiar gaze reinforced by the appropriation and underutilization of technological innovation as ideological reproduction. The private is at once the uniquely other—that experience or struggle conveniently relegated from public concern, the individuated responsibility of isolated, depoliticized subjects—and the scientific, controlling, dissociating subjectivity, "the gaze from nowhere," as Donna Haraway calls it, "tied to militarism, capitalism, colonialism, and male supremacy—[seeking] to distance the knowing subject from everybody and everything in the interest of unfettered power."[60] The public is consequently defined by an abstracted sociality of disembodied, dematerialized, dissociated experiences. In their discussion of "The Sensory Basis of the New Mass Media," Negt and Kluge put it this way: "People are united as individuals, but they experience this union through capital. They can recognize each other only via this apparatus. Collectivities are formed, but without self-regulated interpersonal relationships; forms of satisfaction are developed, albeit passive ones. That which exists is organized, but there is no autonomous activity."[61]

Like the panorama, the satellite is a means of discovery and exploration, of seeing more and seeing more closely, of connectivity and education, exposure and exposition, distance and immediacy, surveillance and control. How these mechanisms and potentials materialize, what remains rhetoric and myth, what serves the powers of exploitation or of mobilization and emancipation depends on the type of relationality, on the quality of the organization of the various "chunks of experience" that is possible and at any given place and moment.[62] In their early 1970s assessment of that quality, Negt and Kluge call the satellite a "symbolic flagpole" of new media in late-capitalist society and its hegemonic distribution of ownership over productive forces and relations.[63] Along with the mass dispersal of hardware, satellites opened new markets for consumption, produced new consumers, *and* made consumable, in turn, the places and people now increasingly available to and, crucially, available through (or *as*) a Western glance from a safe distance—the much-invoked global unity relies on "regulatory forms of communication that do not entail a response."[64]

Such ways of seeing and knowing the world, of navigating socio-geographic territory, of constructing subjectivity in relation to other people and spaces remain "nonpublic in character" as they expand a distanced, disembodied, and depoliticized panoramic view rather than enable "an expansion of the human senses, of our immediate experience, in corresponding to the actual level of human cooperation . . . the *capacity* for perception."[65]

Attempting to model just such an expansion, *Satellite Arts* presented a conception of television as a *technology of assembly*, a form of mass communication that predates electronics and holds its own distinct political potentials. Unlike those that scatter audiences spatially (e.g., conventional broadcasting), temporally (e.g., the Internet), or both (e.g., mechanical printing), assembly is the means of mass communication that unites people in space and time.[66] As Judith Butler has recently contended, in a contemporary era marked by increasing threats to a livable life under prevailing economic and political conditions, and an environment in which the media does not just reflect who the people claim to be but "has entered into the very definition of the people," the gathering of physical bodies can itself be a political act of persistence and resistance. This is not to say that all gatherings are ultimately examples of "democracy in action," but rather that embodied assembly—and, as Butler makes clear, the space of such assembly can certainly be virtual—has an expressive political dimension independent of the expressions that take place during such assembly.[67] Although ostensibly a series of cross-country dance performances—of "continental choreography," as Rabinowitz called it[68]—at its core, *Satellite Arts* posited television as a potential mode of assembly, a technology that could open up spaces for telecommunicative gatherings of bodies, akin to activities such as strikes and public protests.

Hence, the work facilitated both sensory extension and embodied experience. For it to function, participants had to internalize the technology, adapting their perceptions of space and time and recalibrating their bodies' proprioceptive systems so they could operate within the composite televisual world—both to adjust to its peculiarities and to take advantage of the new possibilities it offered (figure 1.7). The artists considered this fundamental to the work, with improvisational dance seen as the ideal activity precisely because it relied on real-time responsiveness and an intense degree of physical coordination, necessitating such bodily

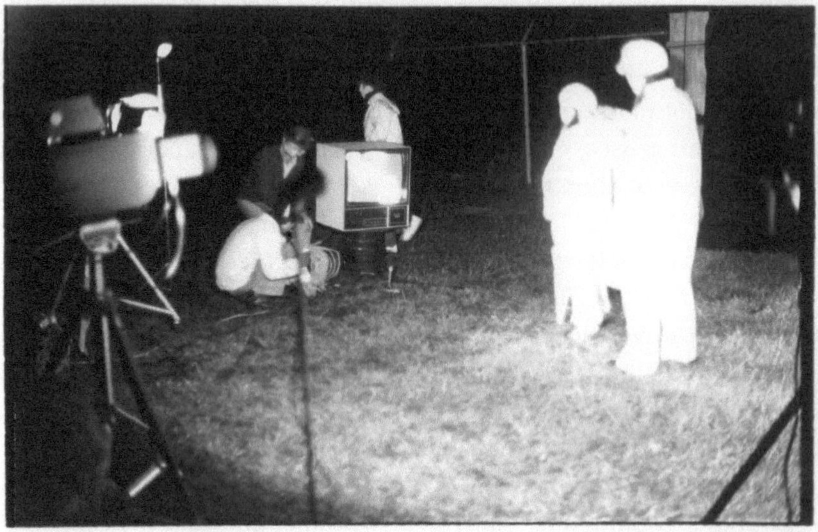

acclimations. The summer prior to the event, Galloway and Rabinowitz themselves performed a series of preliminary experiments by transmitting satellite imagery between NASA Headquarters in Washington, DC, and the Goddard Space Flight Center in Greenbelt, Maryland. Of particular concern was the half-second temporal lag caused by the quarter of a second that it takes for television signals to travel up to a satellite and down to a receiver and the quarter of a second that it takes for the return trip—an effect that had to be mastered in order to make real-time physical interaction possible (figure 1.8). Rabinowitz sat in front of a monitor showing a split screen: on the right was a live image of herself; on the left was a view of the very same monitor she was watching, producing

1.7

Mobile Image, *Satellite Arts (The Image as Place)*, dancers working with monitor, 1977. Courtesy Rabinowitz/Galloway Archive.

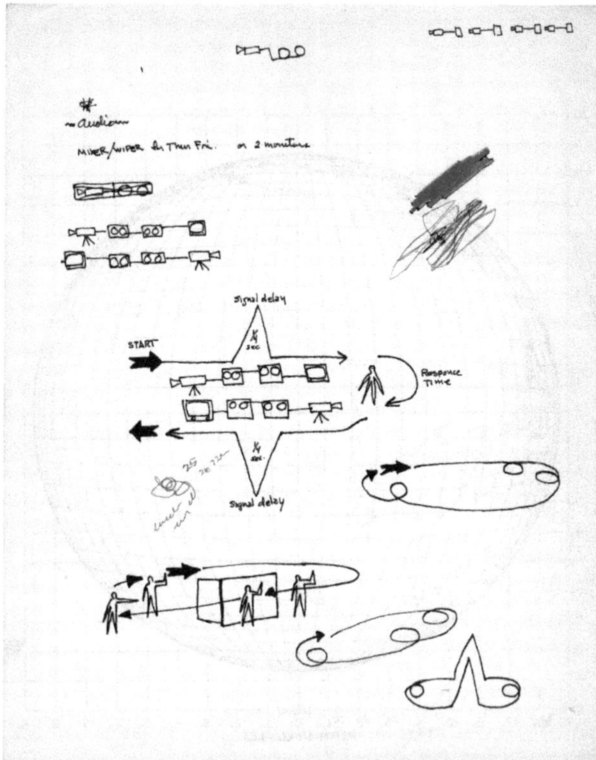

1.8

Mobile Image, notes for *Satellite Arts (The Image as Place)*, showing diagrams with setup and signal delays, 1977. Courtesy Rabinowitz/Galloway Archive.

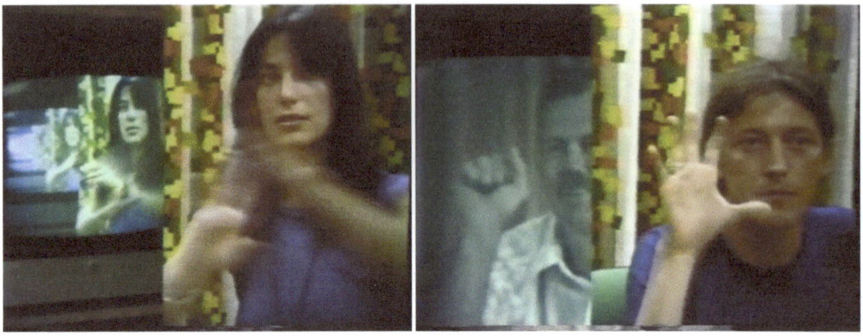

an infinite repetition of her televised visage (figure 1.9). Because of the delay, this repetition progressed in half-second intervals, enabling her to interact with her own image and experience the uncanny sensation of watching her movements travel through satellite space-time. Galloway then attempted to incorporate the delay into some rudimentary physical exchanges with a technician stationed miles away at NASA Headquarters. While Galloway took Rabinowitz's place on the right side of the split screen, the monitor with the technician in it occupied the left side. First, the two played a basic game, in which the technician would make a simple gesture—opening and closing one hand—and Galloway had to mimic it as quickly as possible. The game proved quite difficult. As Paulsen points out, they had to contend with two lags: the satellite delay and the time it took for Galloway to respond after receiving the signal.[69] Next, they attempted to synchronize their actions, requiring them to concentrate on both sides of the monitor, absorbing the visual feedback into their motor coordination in order to align their movements consistently.

1.9

Mobile Image, practice sessions for *Satellite Arts (The Image as Place)*, 1977. Courtesy Rabinowitz/Galloway Archive.

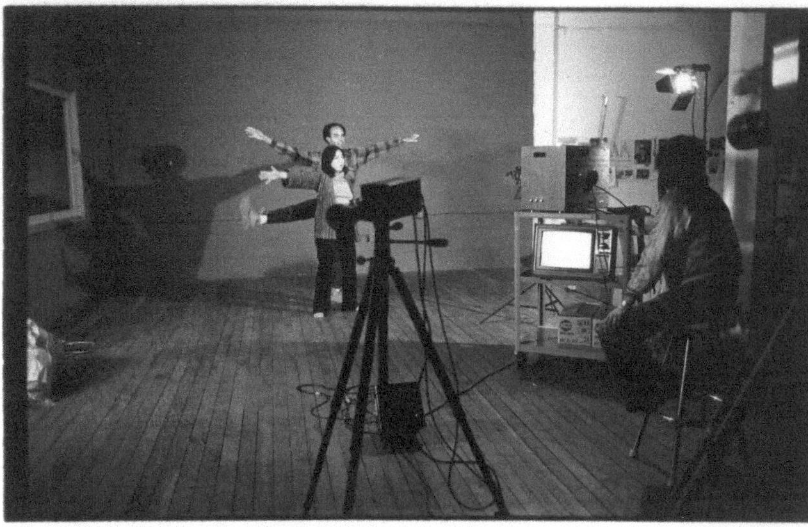

The artists also held a series of rehearsals with the Mobilus dance troupe. Dividing the group among two rooms, they used a closed-circuit video link to simulate a satellite connection so that the performers could begin to acclimate to the televisual space (figure 1.10). It was during these rehearsals that Galloway and Rabinowitz first established the composite screen in which bodies could assemble and act together (figures 1.11 and 1.12). They preferred this "visual architecture," as Rabinowitz called it, to the split-screen format, which they saw as simply connecting two geographic points for an exchange of information, maintaining unwelcome divisions between participants. (In Crary's terms, the split screen "transcended local provincial viewpoints" but still preserved "subjective isolation, . . . sensory impoverishment and emotional privatization.") The technical setup therefore had to be customized to facilitate a place for

1.10

Mobile Image, rehearsal for *Satellite Arts (The Image as Place)*, 1977. Courtesy Rabinowitz/Galloway Archive.

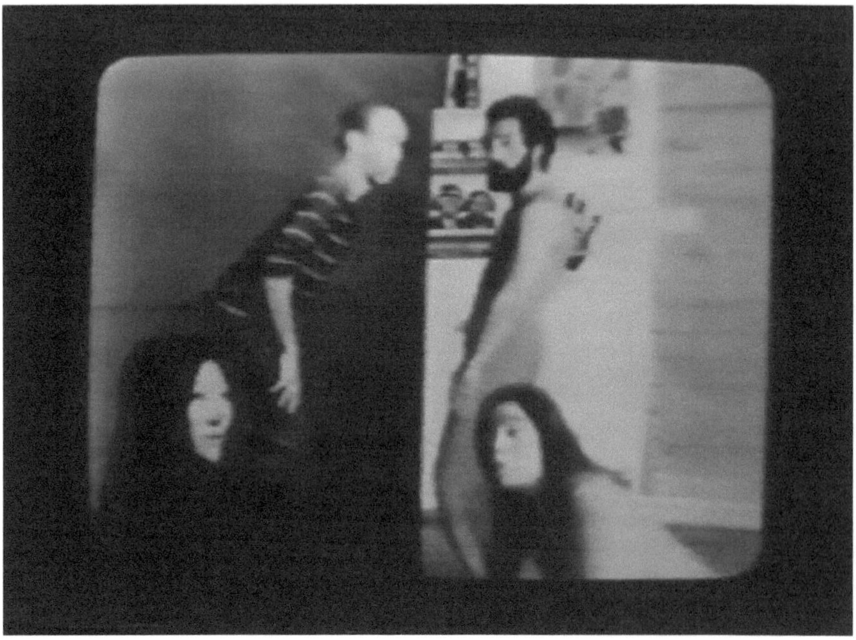

assembling. The monitors that the dancers used to coordinate their movements had to be shown in mirror image, for instance, so that leftward movements appeared to go left and rightward movements went right. And the precise boundaries of the performance spaces had to be defined in relation to the cameras, all of which had to be set at the same height so that the performing bodies would line up properly.[70] The rehearsals were limited, though, because the biggest obstacle to performing in a composite screen—the satellite delay—could not be simulated in a closed-circuit setup. *Satellite Arts* therefore began with a brief "warm-up" phase, just prior to the commencement of the scheduled series of dances. As they moved through the virtual environment, familiarizing themselves with its parameters and making room for each other, the performers began

1.11

Mobile Image, rehearsal for *Satellite Arts (The Image as Place)*, 1977. Courtesy Rabinowitz/Galloway Archive.

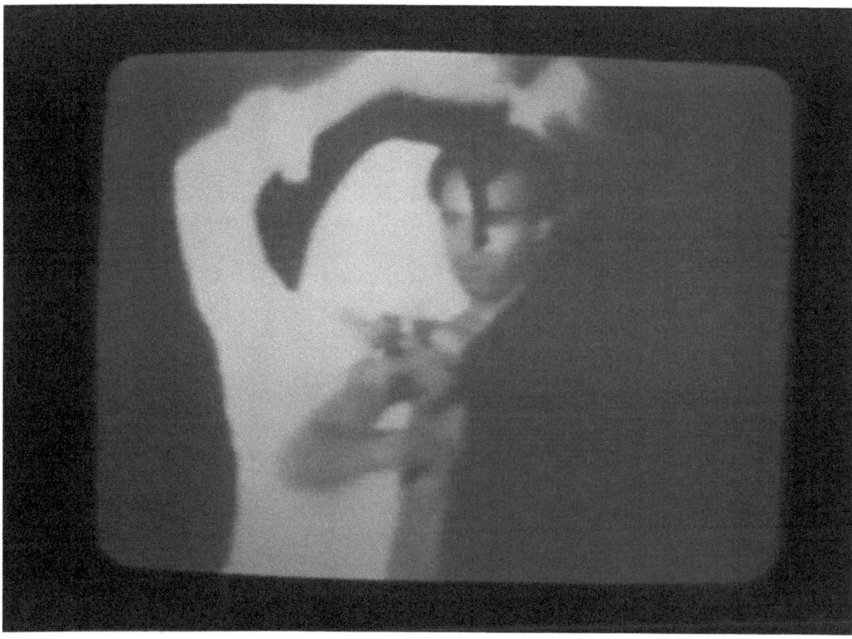

to experience a new spatial, temporal, and corporeal existence, distinct from both the "real" world and the mesmerizing effects of conventional television. Rabinowitz explained this phenomenon:

They had to adapt to the disembodied reality of that thin, two-dimensional space, yet their interactions had a "thickness" determined by the satellite time delay, a kind of molasses movement. It was thin space/thick time, like an out-of-body experience, the feeling of a transcendent dream. You give priority to your image as a kind of ambassador in virtual space. But even though you're separated from your body it's incredibly sensual. The sensory deprivation actually enhances sensitivity, heightens your appreciation of sensual experience. You "own your own

1.12

Mobile Image, rehearsal for *Satellite Arts (The Image as Place)*, 1977. Courtesy Rabinowitz/Galloway Archive.

image" so completely in real time that it's like having phantom limbs. After living in that space a while you have to reorient back to bodily sensations.[71]

Ultimately, the extension of oneself into the televisual was, paradoxically enough, both real and simulated, representational and experiential, disembodied and resolutely bodily. Rabinowitz continued:

How do you describe the experiences of dancing a half-second in the past, in real time, live with someone who's not really there, on a stage that exists only as a 2-dimensional image, coordinating a self that is outside of your body, where all contact and communication is electronically mediated, where performers are spectators at their own performance, having the same visual perspective of the rest of the audience, in a complex conceptual reality that exists only on a television? The Satellite Arts Project was a hybrid experience producing a hybrid vocabulary to describe phenomenon like the "disembodied self," "phantom space," "split-screen schizophrenia."[72]

The power of the televisual is its extension inward and outward: it offers greater visibility and vision, participation and access to information, sites, and experiences, while at the same time exerting control over those people and things it frames and transmits, and over those who perceptions are reduced to ostensibly subjective and diverse yet predetermined choices. Television, as Richard Serra so astutely observed, not only packages the outside world into agreeable objects but also, in turn, "delivers people."[73] Or, as Vilém Flusser writes: "The box has buttons which offer the viewers the choice of various channels, and can also interrupt the flux of the messages. This creates an impression of control over the box and of a sort of mechanical freedom."[74] The televisual lets the viewer choose, participate, and interact, while curbing the machine's communicative potential: "The box emits messages but does not receive any."[75] It is important to reemphasize that the political and socioeconomic power of media cartels and communication monopolies was understood to be exercised not simply through the inescapable imposition of unilateral commands but also via an appeal to cultural personalization and choice, the interpellation of subjects as multiple, free, individuated citizen-consumers. Rather than "fear the masses," television and new media, Enzensberger explains, "tendentially do away with all educational privileges," not by homogenizing all broadcasting content but conversely through the diversification and dehierarchization of available information.[76] Rather than violently impose and exploit false needs, the "consciousness-industry" proceeds with "the falsification and exploitation of real and legitimate ones," without which it would be ineffective. These include "old psychosocial themes" such as social prestige and identity, as well as powerful new "collective wishes"—nonmaterial variety, mobility, participation in social process, new forms of interaction, release from ignorance and tutelage, self-determination.[77] Connectivity and transmission as such become the determining factors of the relationality we construct between isolated segments of experience. In her prologue to *The Logic of Television*, Patricia Mellencamp writes that "John Berger's assessment in 1972 of the replacement of political choice by a polynomial begetting of new and improved products is right on the money . . . Amidst this infinitesimal *differentiation*, the political concept of *differences*, along with meaningful choices, is arduous to maintain."[78] "Localized diversity" propels the myth of the global village, substituting individuated connectivity for a politicized and perceptible relationality of social and economic, racial, gender, and other different forms of experiences, or experiences of difference, conspicuously absent from news, soap operas, and talk shows.

Under the televisual regime, the contradiction between productive forces and relations of the televisual is exacerbated by new media innovation, such as, during the 1960s and 1970s, the reach and immediacy of the satellite, the customization and variety of cable access, and the personalization and mobility of the video camera. More information, and an increasingly "free flow" of it, blurred the boundaries between public and private as an endless barrage of televised images became ubiquitous in US living rooms, kitchens, and bedrooms. At the same time, the televisual cemented the notion of the private as the site of domestic consumption rather than the locus of the social production of subjective experience, where viewers would interpret, negotiate, and contest their relationship to the world around them in a communicative sense, hence, as Miriam Hansen puts it, in an "intersubjective, potentially collective, and oppositional form."[79] The viewer remains a user, reaching out and creating relationalities in stunted, unidirectional ways, exteriorizing a dematerialized body through technology as a dissociated observer experiencing privatized, distracted, fragmented encounters. The relation between inside and outside, self and other, local and global, private and public goes both ways. But it is not *communicative*, as it fails to establish conscious and critical, productive relations between the various outsides and among the various insides, further rendering public and private as monolithic, rather than dynamic and political, entities. Under such circumstances, the demand to make media and technology *more* public by expanding audiences and access would entail restricting communication to formal, technical conditions such as free association and equal participation, while excluding that which, as Hansen explains, "the dominant public sphere either leaves out, privatizes, or acknowledges only in an abstract and fragmented form." These exclusions include: experiences of alienation produced by the discrepancy between what Enzensberger calls "real and legitimate" needs and their only partial or distorted, passive gratification; the meaningful connection of these experiences to other ideas, fantasies, and memories; and the refunctioning of available tools to exchange and ultimately organize these experiences and connections, to construct new social formations of interrelated and competing publics.[80] Without such productive and critical relationality, technical innovation is bound to reproduce, and increase, technology's hegemonic function.

Satellite Arts can be understood as an attempt to foster an experience of this relationality by establishing a time and a place for reciprocal telecommunication, by modeling a state in which the desires for individuality and difference, on the one hand, and commonality, on the other, were activated in all their seemingly incompatible interconnectedness. As Rabinowitz's comments suggest, the "out-of-body experience" of the televisual world generated its own communicative sensibility, a unique type of electronically mediated connection by which one became part of a collaborative whole, outside of themselves—a form of "intersubjectivity." As Bruce Clarke explains, intersubjective communications "effectively render exterior and collective what would otherwise remain interior and private, as the self-knowledge of an individual subject enters the public sphere. Such communication would entail the merger or intermingling of inner and outer, psychological and social events."[81] Yet, the challenges for the artists were to articulate this "merger" as emphatically contradictory and never complete, and to prevent the symbolic erasure of individual subject positions through the contrived bliss of virtual union. *Satellite Arts* staged a conception of social subjectivity as temporary, fragile, and continuously renegotiated. Without its reflexive, theatricalized performance of contradictory intersubjectivity, the work's valorization of collective identity and apparent transcendence of individuality would have risked eliding existing experiences and inscriptions of difference. Not everyone benefits from such elisions—indeed, they arguably reinscribe power relations.

With *Satellite Arts*, Mobile Image focused on fulfilling the expansive possibilities of ubiquitous but underutilized telecommunication systems, specifically attempting to transform television into a technology of assembly, an "image as place," where people could gather and act together. As Enzensberger explained in 1970, "The open secret of the electronic media, the decisive political factor, which has been waiting, suppressed or crippled, for its moment to come, is their mobilizing power."[82] As opposed to the dematerialized and disembodied—and thus privatized, isolating, sensorially impoverished, and ultimately depoliticized—mobility promised by the televisual regime, this mobilization is physical and psychological, affective and intellectual, political and social. *Satellite Arts* reflected an ambivalent understanding of the media as untapped potentiality rather than as always already part of a totalizing and condemnable "culture industry." Yet, the work was not simply about creating more opportunities for individual expression or turning the tables on power, attempting to redirect controlling means to "better" ends or forming media subcultures composed of alternative, parallel channels of broadcasting. As Enzensberger contends, "Anyone who imagines that freedom for the media will be established if only everyone is busy transmitting and receiving is the dupe of a liberalism that, decked out in contemporary colors, merely peddles the faded concepts of a preordained harmony of social interests."[83] Galloway and Rabinowitz similarly recognized that only new organizational models, new modes of what they called "being-in-the-world," beyond the broadcast, can be politically effective. *Satellite Arts* restructured the television apparatus into a way to make people more mobile, "as free as dancers, as aware as football players, as surprising as guerillas," to use Enzensberger's words, thus "bring[ing] the communications media, which up to now have not deserved the name, into their own."[84]

Mobile Image attempted to reorganize radically the relationality among and between subjects, between self, other, and world. The artists saw the project as an attempt to create new spatial and social mobility, to create what Galloway and Rabinowitz called "a new 'scaled perspective,'" one that would redistribute sights and sites thus far contained by and within the logic of the panoramic, of the televisual, and its ideologically and physically prescribed perceptual positions. "The key dimension of our time," they contended, "is the scale that separates technological possibilities from human imagination and understanding. The challenge—for the artist, for us all—is to reconcile the dialectical relationship between quantitative technology and the qualitative desires of humanity."[85] Echoing Enzensberger's remarks regarding the contradictions between productive forces and relations, *Satellite Arts* aimed to transform the apparatus, not simply its devices, but their very function and utility as part of the production and reproduction of social and ideological institutions. As if to address Schiller's concern about the imperialist and colonizing mechanisms of new media directly, whether on a domestic or international level, Mobile Image sought a fundamental repositioning of the (Western, bourgeois) subject as perceiving and producing self in public, in both physical and psychosocial space. As mentioned, Galloway and Rabinowitz recognized from the outset that a communications revolution is primarily about establishing alternative social realms as "laboratories of resocialization, as technologies of the self."[86]

In many ways, *Satellite Arts* recalled what Tom Gunning terms the "cinema of attractions," an early twentieth-century cinema experience in which "theatrical display dominates over narrative absorption."[87] Just as the history of television, and telecommunications in general, has been subsumed under the logic of broadcasting, "the history of early cinema, like the history of cinema generally," Gunning explains, "has been written and theorized under the hegemony of narrative films."[88] Yet, early film offered a competing model, an "exhibitionist cinema" that foregrounded the technology's role as purveyor of tricks and magical effects, its very ability to show something. A cross between theatrical machine and amusement park, it created a specific relation between the projected world of the apparatus and the perceiving spectator, "a cinema that displays its visibility, willing to rupture a self-enclosed fictional world for a chance to solicit the attention of the spectator."[89]

Less concerned with undermining narrative absorption, *Satellite Arts* sought to link spaces, that of the mediated and the mediating, as well as the tool that connected the two in reciprocal dependency. The machine, so to speak, shows itself as a connector of spaces, shows itself showing, as a mechanism of orientation and points of view, of place and site making. The work presented telecommunications technology as a particular mediation and manipulation. As a "TV of attractions," it established a jolting, self-conscious experience of television, by and through which the subject is *positioned* as a panoramic observer in a particular place at a particular time, glancing at space outside the private, as a self in the televisual. It politicized the machine—camera, screen, and satellite—as that which institutionalizes the comfortably contained space of centered perspective.

Participation in *Satellite Arts* was therefore a matter not of breaking free from the television apparatus but rather of always recognizing one's relationship to it and place within it, in order to transcend by refunctioning it. The work's "scores" generated experiences of technological criticality and self-awareness from the start. The initial transcontinental contact between the performers happened in a split-screen format, with West Coast dancers Keija Kimura and Soto Hoffman on the left and East Coast dancers Nathan Stinson and Mitsuko Mitsueda on the right of the vertically divided screen (figure 1.13). As they excitedly waved to each other, the first views they saw of themselves and their partners were at oblique angles, not from the cameras facing them head-on but rather from ones set off to the sides to document the setup and process from a distance. The dancers thus immediately saw themselves as part of a larger arrangement, as nodes in a network of people and parts—backdrops and flooring, on-site technicians and camera-people, microphones, monitors, and the large television cameras pointed at them. Once the connection was established, both groups discussed the physical arrangement of the space, its boundaries, the audiovisual devices, and how these components aligned with the televisual world they were preparing to enter. They focused on the structure and limitations of the bisected screen on which they and their collaborators appeared, gesturing toward one another, shouting,

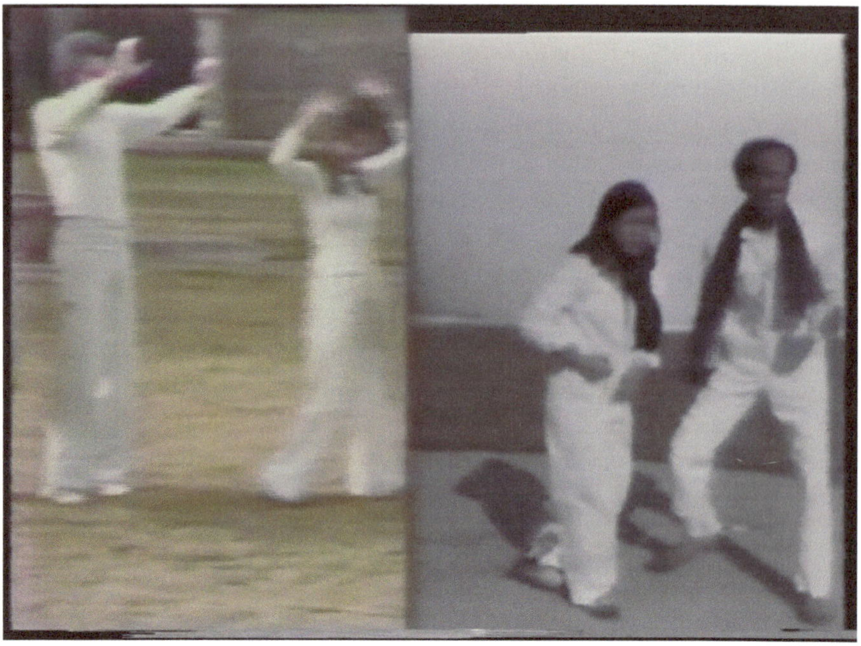

and running their hands along the screen's vertical "seam," whose stark border severed any body parts that extended beyond it. This initial activity underscored the physical distance between the groups. "The line divide[d] the image and constrain[ed] the dancers," Paulsen explains, "and in doing so it accurately diagram[ed] the technological situation: two video feeds from opposite end of the country occupying opposite ends of a television monitor."[90]

The dissolution of that line was dramatic, as a much more unusual composite screen space suddenly came into being, traversable by all participants,

1.13

Mobile Image, *Satellite Arts (The Image as Place)*, video still, 1977. Courtesy Rabinowitz/Galloway Archive.

regardless of their actual locations. "Precision Dance" was the first, foundational score (figure 1.14), consisting of the performers cycling through twelve basic arm gestures so that they could begin to synchronize their movements. Although this score eventually transitioned to an outdoor setting, it opened with the dancers superimposed on a live view of the control room, whose monitors showed the premixed feeds of the dancers in each location (figure 1.15). The performers thus saw themselves overlaid on a scene of the production team and equipment processing those live feeds. In these ways, the very first tests of televisual communication deliberately incorporated the structure of their constitutive illusion. The effect was distinct from that of contemporaneous works such as Michael Asher's control room broadcast, which spotlighted the material conditions of television by showing its backstage operations, but still functioned as a program transmitted for TV-viewer reception. *Satellite Arts*, by contrast, repositioned such viewers—now no longer really viewers at all but rather participants—inside the apparatus, affording them an experience that was decidedly less behind-the-scenes than, as Galloway and Rabinowitz put it, "within-the-scenes."[91] Everyone involved—dancers and technicians alike—was immediately made palpably conscious of the technological apparatus that would contain them, enable their subsequent communications, and determine their relationships to each other and to the technology being used.

Other tele-revolutionary efforts of the early to mid-1970s similarly attempted to generate this kind of palpability, this theoretical-reflective tangibility of the apparatus as mediated and mediation, as conceptual mechanism, social construct, and technical device. Rather than re-entrenching traditional dichotomies of the real and the virtual, experience and information, authentic and manipulated, guerilla television groups such as TVTV sought to materialize the televisual image through the vivid occupation and activation of the spaces in front of the camera and, by extension, in front of the screen. In 1971, Michael Shamberg and the Raindance Corporation video collective demanded a radial "redesign" of the "ecology" of "Media-America."[92] One strategy was to connect ostensibly disparate sites of production and reception in order to create the physical and psychological "feedback" needed for the "verification of experience," the very prerequisite for social progress.[93] In 1972,

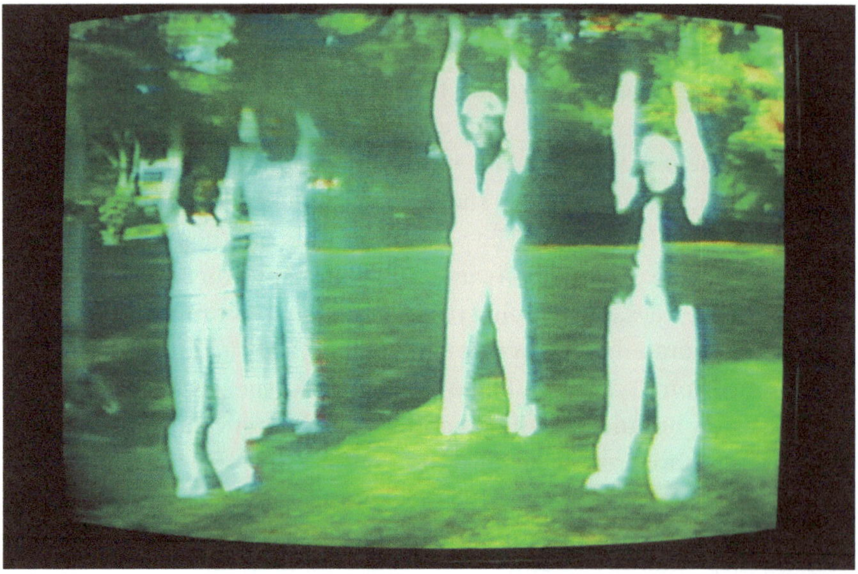

Shamberg joined members of the art and media groups Ant Farm, Videofreex, and Raindance to form TVTV's first production team. With the support of four cable television stations, the crew descended upon, first, the Democratic and, a month later, the Republican National Convention. Footage from the latter was compiled into an hour-long video tape titled *Four More Years*, subsequently shown on various CATV channels throughout the United States. The video shows TVTV's nineteen-member team making its way through a range of events—delegate caucuses, Young Republican rallies, cocktail parties, antiwar demonstrations, the chaos of the convention floor—capturing the frenzy of zealots, politicians, and the press.[94] The tape denies both producers and audience the comfort of ostensibly "objective" distance, avoiding voice-over commentary and added music and instead constructing a vivid collage: campaign-button

1.14

Mobile Image, "Precision Dance," *Satellite Arts (The Image as Place)*, video still, 1977. Courtesy Rabinowitz/Galloway Archive.

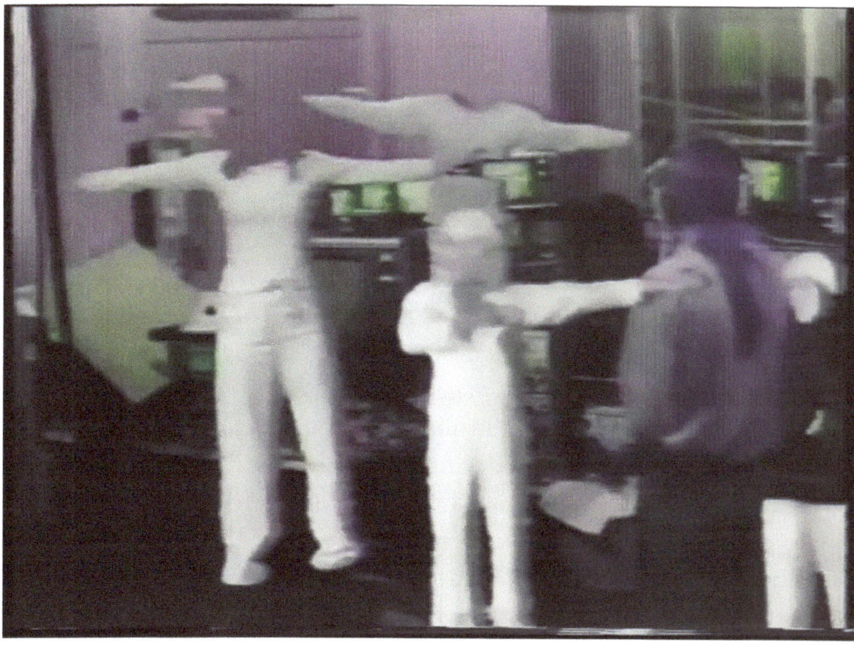

close-ups, portable cameras pointing at big TV crews and away from podiums, protest sound bites, off-the-cuff moments with politicians and their families, and behind-the-scenes conversations with network personalities such as Walter Cronkite and Mike Wallace. Deirdre Boyle likens the documentary style of *Four More Years* to "*cinema vérité* of the 1960s," shot "from *within* the crowd, subjective and involved," as the small, unobtrusive size and light weight of the Portapak cameras afforded the crew as well as the viewers a nimble and immersed navigation of events.[95] Efforts such as these constructed a playful, perplexing, yet analytical relationality between points of view as the broadcast industry itself and the viewing positions and spaces it produced became subjects of television.

1.15

Mobile Image, "Precision Dance," *Satellite Arts (The Image as Place)*, video still, 1977. Courtesy Rabinowitz/Galloway Archive.

Boyle contends that TVTV's "rise and fall traces a major arc in guerilla television's history." The group's efforts came to an end in the mid to late 1970s amid what, in the context of Mobile Image's attempts to refunction new media's televisual logic, could be called a renewed imbalance between productive forces and relations, with TVTV's later concerns privileging the latter. The group's focus evolved from an interest in the politics of broadcasting to its de facto extension. Rather than further expanding telecommunicative spaces of production, wherein audiences not only began to understand themselves in their roles of stunted mediators and producers of experience but also would wield the tools of new media in emancipatory ways, *as means of new social production*, the collective became increasingly involved in the quantitative expansion of distribution. Conventional relationships between producers, audiences, and content were reestablished, along with what Boyle calls "a slicker, TV look."[96]

Coming on the heels of this reversion, *Satellite Arts* sought to transform the quality of televisual experience rather than improving production and increasing the quantity of distribution. Mobile Image was in many ways the beneficiary of the labor and experiences provided by TV and video activism. Yet, *Satellite Arts* challenged distinctions between performer and technician, process and product, artist, engineer, and audience. The work was conceived as a laboratory for experimentation, as a "research and development" venture, the primary experience of which was the negotiation of the technical arrangement, the processes of managing both the limits and new possibilities of that arrangement, of acting within a matrix of mediators and mediation, and, just as crucially, of reporting back on the findings. The scores were devised as improvisations in order to make this negotiation the central focus. The rehearsals could only provide so much preparation:

> We had become a tight group who could work and problem solve together, but we had no idea what choreographing the entire event via satellite would be like. Though we had performed in visually separated spaces during the entire rehearsal period, the truth of the matter was that we relied upon face-to-face gathering around the table for discussion, problem solving and brainstorming. We had no experience in doing everything, in communicating totally through the linked television sets . . . The satellite project would test our collective ability to

choreograph and perform all aspects of the performance via the satellite linked TV window.[97]

Everything had to be managed through the apparatus—the work's complex logistics, along with the myriad problems that arose—and all communications had to occur over the same audiovisual channels. Technical comments, instructions, and reactions were by necessity incorporated into the scores, producing a multidirectional feedback loop that enabled participants to consider, respond to, and adjust whatever role they were performing within the system continuously.

Satellite Arts thus generated a critical consciousness of techno-social relations, not just by pulling back the curtain, so to speak, and revealing the machinery—control room, monitors, cameras—but also by foregrounding the apparatus in ways that forced each participant to confront the mediation of his or her experiences, the various agents—human and mechanical—involved in that mediation, and his or her position vis-à-vis that arrangement. Nearly all of the scores did this in one way or another. "Triangle Dance," for instance, consisted of Kimura, Hoffman, and Mitsueda performing together in an open field, with Mitsueda projected into the others' locale (figure 1.16). "The score required that the dancers be constantly focused on the monitor, choreographing and placing themselves, being aware of the camera movement in relation to the illusionary 'edge' of their partner's image," Galloway and Rabinowitz explain.[98] At one point, the West Coast camera zoomed in on Kimura and Hoffman, rendering Mitsueda's body comparatively tiny. The dancers had to accommodate this shift in scale spontaneously, with Mitsueda suddenly contending with giants, and the others interacting with a miniature figure who at times appeared to climb and dance on their enlarged bodies. As Galloway and Rabinowitz recalled, "These variables of camera movement and zoom combined with the movement of the dancers multiplied the possibilities of improvisation and graphic interpretation and demonstrated the interactive involvement not only between dancers but between camera man and dancer and camera man and camera man."[99] "Human Time Delay Feedback" involved two of the dancers—Kimura on the West Coast and Stinson on the East Coast—attempting to mimic each other's movements and speech. The dialogue immediately became a running commentary on

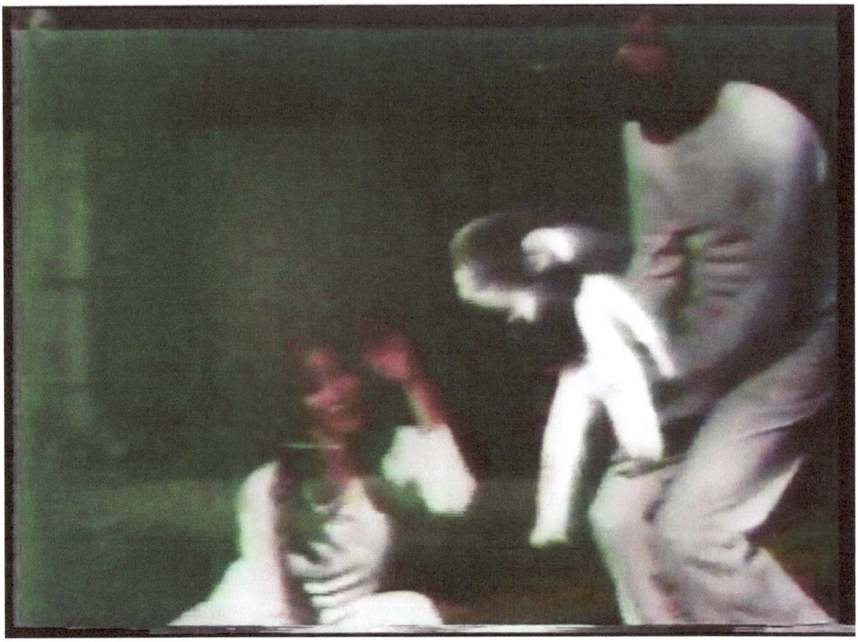

the communications structure itself, particularly the technical difficulties emerging at that moment: "The East Coast has no audio"; "Can you see me?"; "Can you hear me?"; "Am I coming through?"; "Are you listening?" "Process Dance" was meant to build upon the first day's experiments, requiring participants to engage the technology more emphatically and self-consciously now that they had sufficiently acclimated to the setup. In this score, planned but not executed because of the difficulties on days 2 and 3, the dancers were to contemplate openly that setup, its challenges and processes of "problem solving and decision making," and the fact that, as Galloway and Rabinowitz explain, "making the system work

1.16

Mobile Image, "Triangle Dance," *Satellite Arts (The Image as Place)*, video still, 1977. Courtesy Rabinowitz/Galloway Archive.

was part of the choreography." "Process Dance" was again to incorporate control-room footage into its score, so that the dancers could respond to and interact with the means of their own mediation.[100] Such components of *Satellite Arts* became, as Galloway and Rabinowitz put it, a "self-reflective commentary on the process of the performance."[101]

This experience of relationality also extended outside the contained network established by the work's configuration, equipment, and the interactions of its participants. The overall project—its technological arrangement and the set of relations and experiences established by it—was itself set in relation to its broader context, both the dominant media environment and the specific geographic sites ostensibly overcome by the creation of a third space in the screen. The extraordinary experience of extending oneself across thousands of miles at the speed of light was one not of pure transcendence of time and place but rather of paradoxically being in two places and two times at once. These moments of telecommunication, the apparent elimination of distance, also conversely underscored the participants' "real world" positions, the actual distance between them, and the differing experiences produced by such positioning. This contradictory effect was largely the product of the technological arrangement itself—the fact that the participants had to remain cognizant of their own studio space and the televisual space simultaneously, along with the pervasive, but as it turns out quite varied, time delay. Because the main control room (figure 1.17) was situated where the West Coast dancers were performing (the Educational TV Center in Menlo Park), the feed was hardwired there, meaning that those dancers actually saw themselves with no delay. On the East Coast, however, the camera and microphone signals had to be transmitted to the satellite (with a delay of an eighth of a second) and then relayed from the satellite to the control room (another delay of an eighth of second), where both feeds were instantaneously mixed into the composite image and transmitted to the West Coast monitors (no delay) and via satellite to the East Coast monitors (a delay of a quarter of a second). This resulted in very different audiovisual experiences, depending on where each participant was physically located: the West Coast dancers saw themselves in real time and their partners in quarter-second delay; the East Coast dancers saw themselves in half-second delay and their partners in quarter-second delay. Meanwhile, everyone contended

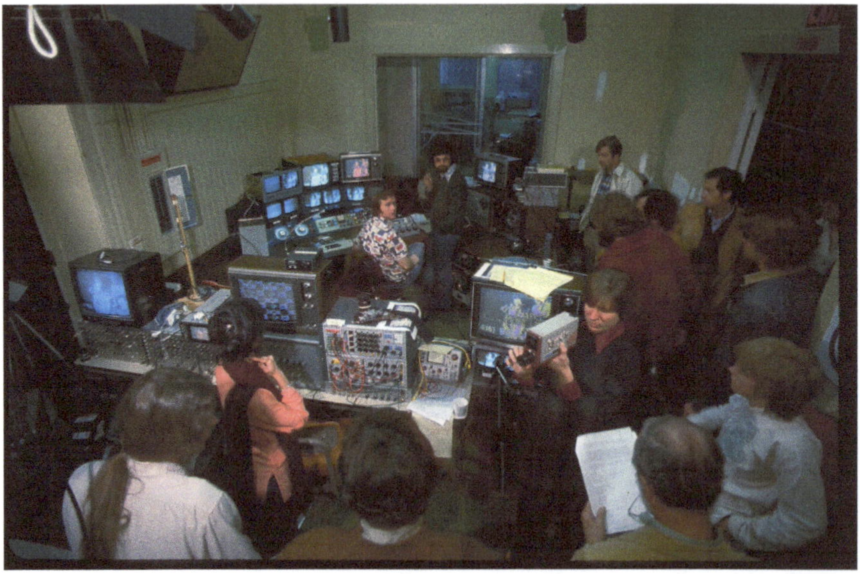

with various degrees of echoing caused by sounds traveling through the system in a continuous circuit. Because the performers had to both transmit audio and hear that of their cross-country partners, the microphones and speakers had to face the same direction, producing inevitable feedback loops that took on unique qualities because of the time delay. Rather than suppressing such feedback, the control-room operators maintained it by continuously monitoring and balancing it. As indexes of the actual distances being traversed, these visual and aural effects conflicted with the satellite's supposedly unifying function, rendering such distances palpable. Participants became abundantly aware of how far they were from each other and from the equipment being used. Mediation was, in this sense, materialized. As the East Coast camera operator Steve Christiansen

1.17

Control Room, *Satellite Arts (The Image as Place)*, 1977. Courtesy Rabinowitz/Galloway Archive.

put it with regards to the unusual audio effects, "Our sense of hearing was extended in a way that *made me feel the distance of the satellite and the acoustic space it creates*."[102]

Satellite Arts also expanded the relational field to include the wider world of mass media television. In the "Real Time TV" score, for example, the four dancers were placed within a live television feed, taken directly from whatever was being broadcast at the moment, determined by someone in the control room simply flipping channels. This score came during the end of day 1, after the dancers had been experimenting with the arrangement for some time. Now, that arrangement was suddenly complicated by the intrusion of the live broadcast as the dancers experienced the disorienting sensation of being at once watchers of TV programming and actors within that programming, over which they themselves had no control. Whoever was controlling the dial ultimately landed on a live football game between the Los Angeles Rams and the San Francisco Forty-Niners—a choice that would have also underscored the West Coast source of the broadcast (figure 1.18). The dancers attempted to "join" the game, mimicking the movements of the athletes and jumping up and down when they scored. (The camera operators had to respond to both the dancers and the broadcast continuously: as the latter shifted perspectives, the zoom had to be adjusted to keep the dancers' bodies in proportion with the players.) At one point, the West Coast camera pulled back to reveal two people holding the black felt curtain behind the dancers, the surface on which the television broadcasts were keyed. The football game still appeared there in the composite image, but the illusion was shattered by the expanded frame, which now included the actual grassy field in which the West Coast dancers were performing.

As with Brecht's *Lehrstücke* (or "learning plays"), *Satellite Arts* attempted not to "activate" the viewer but rather to model a different technology of the self—or an understanding of the self as technological. Developed in conjunction with his "radio theory," Brecht's learning plays had no audience, illustrating the way in which the playwright thought of the function of politicized aesthetics in general. Rather than thinking of information, knowledge, and habits bestowed onto and guiding the viewer from the outside, any revolutionary artistic practice had to include the experience of alienation: the relation or constituting and

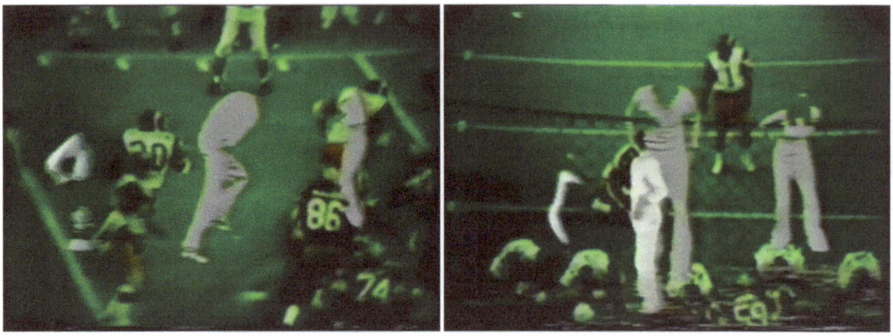

contradictory dynamic between what is and what ought to be, between private and public, here and there. For Brecht, the goals were for a collective to examine its own conditions and to chart the parameters of its possibilities for change: "The *Lehrstück* teaches by being played, not by being seen."[103] The notion of ownership (*Eigentum*) is extended to the production of experiences, to "selfhood," and Brecht's goal for the learning play is the formation of *Eigensinn*, a "sense of self" as historically and technologically determined and determining subject.[104]

Here, the distance between stage and audience is neither collapsed nor abandoned altogether for a more authentically "participatory" space. Rather, it is articulated and refunctioned, performed as an in-between space, as the site of multiple connections between and among various sites of production. Building upon Mobile Image's declaration that *Satellite Arts* establishes a "third space," a meeting ground or "image as place," Paulsen argues that the artists "destroy" the dualities of here and there, subject and object, then and now. "By crafting a mixed image space from multiple camera feeds," she explains, "the artists used screen space to model an impossible and idealized phenomenological situation in which

1.18

Mobile Image, "Real Time TV," *Satellite Arts (The Image as Place)*, video stills, 1977. Courtesy Rabinowitz/Galloway Archive.

the binary differences that govern our typical experience of the world dissolve."[105] Yet, the work's significance does not end at this dissolution. *Satellite Arts*' most crucial contribution to a theory and practice of new media is that it models not a third space, a discrete and dematerialized site of utopian suspension, but rather a *constellation*, a process, a set of relations and negotiations that is as mediated, real, and material as that place opposite the camera, in the screen, up in the sky, and in front of the screen. The dancer's image on the monitor becomes not just an "avatar" or "ambassador" but also the subject as technological self, always already mediated. The sensible dis- and reorientation provoked by a television of attractions, the "stimulated unhealthy nervousness" that Gunning sees at the root of the historical avant-garde's desire to "organize popular energy for radical purpose," formed the basis for *Satellite Arts*' laboratory of resocialization.[106]

Thus, while the work modeled a radically new conception of televisual communication, the spaces it established were never self-contained, never autonomous, always set in relation to other spaces, other devices, other people, the apparatus, the world. Denying the spectator position of the panoramic observer, who is typically locked within his or her position as a seemingly freely roving yet thoroughly constrained (and unself-conscious) eye, *Satellite Arts* essentially articulated a dialectics of the televisual: as with Brecht's "learning plays," it was less an attempt to activate spectatorship as an alternative to "narrative absorption" than a counter-vision of the "image as place" as a process space, as flux, as negotiation. Empowered to extend themselves and occupy that space, to improvise and collaborate across an enormous distance, participants were also reminded that they were part of an elaborate technological arrangement, that they had to share control of their experience, and that each collaborator was set within a larger relational field, simultaneously both an actor and a spectator, a sender and a receiver, an observer and one being observed. Recognizing the severe limitations of the standard broadcast logic, *Satellite Arts* did not simply invert it, turning television watchers into participants, consumers into producers. In fact, the work overcame the active–passive binary at the core of both that logic and its inversion. As implied by this work, "activation" is not just about mobilizing spectatorship, which arguably already happens in conventional mass media, from the visitor who travels to Paris via a panorama, wandering around taking in its views, to the couch potato who can go virtually anywhere by flipping through hundreds of channels from the comfort of a living room sofa. As Jacques Rancière has recently argued: "We don't need to turn spectators into actors. We do need to acknowledge that any spectator is already an actor of his own story and that every actor is in turn the spectator of the same kind of story." "Emancipation" entails "a reframing of the very relationship between *doing*, *seeing*, and *saying* . . . the blurring of the opposition between those who do and those who act, between those who are individuals and those who are members of a collective body," in order to recognize and potentially reorganize "the given distribution of the sensible."[107] *Satellite Arts* modeled precisely this process of reframing, blurring, recognition, and reorganization.

EMBODIMENT

Before cofounding Mobile Image, Rabinowitz studied architecture at the University of California at Berkeley, and from early on, she considered the engagement with the material and immaterial qualities of interactive environments to be a crucial intersection of art and politics. In 1972, she joined the photography and soon-to-be-video collective Optic Nerve (figure 1.19), which was part of Project One, an early example of warehouse communities in San Francisco's Mission District consisting of dozens of artists, media, and political activists. Optic Nerve was no stranger to collaborative efforts and, by the mid-1970s, had established itself as an integral component of the alternative video movement in the United States, an informal but closely connected network of initiatives that Gene Youngblood referred to as "a community of consciousness."[108] In the Bay Area, this community also included Video Free America, TVTV, and Ant Farm. "Collectivism was a live-style of the times," media scholar Marita Sturken explains in a text published in a 1990 Bay Area Video Coalition anthology. "The prevalent ideology was one of sharing—living environments, work, information."[109] In her contribution to the same publication, media scholar Deirdre Boyle adds: "Community video activists were not only dedicated to serving regional constituencies but also

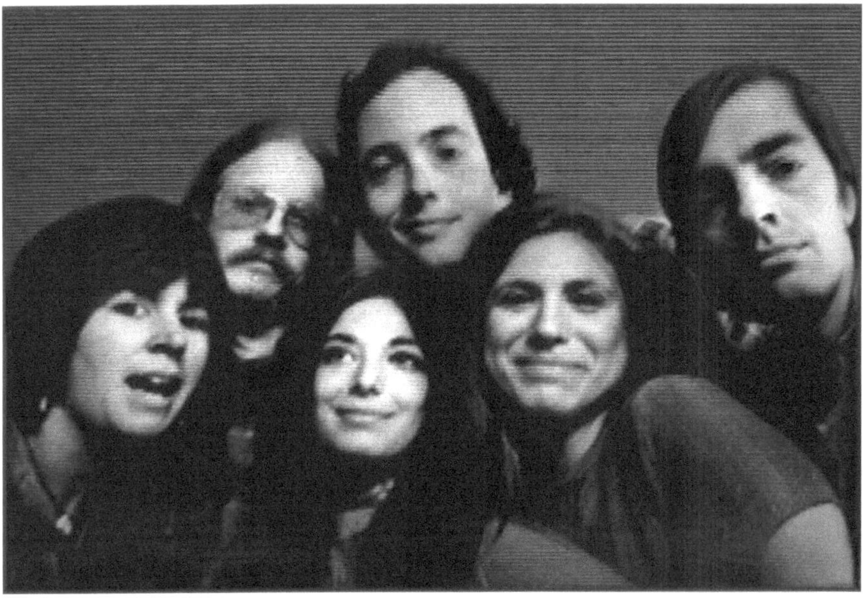

to serving the specialized interests of multicultural communities such as women, gays, blacks, Latinos, Asians, and Native Americans, among others."[110] Boyle points out the concurrence of the debut of portable video and the beginning of the women's movement, the innovative technical apparatuses granting access to technologies of storytelling, communication, and broadcasting: "Women exchanged video letters; they started their own video access centers and programmed their own cable channels; and they ran their own video festivals."[111] Optic Nerve was a crucial part of this history, producing "the first feminist documentary" in 1973 on the Miss California Beauty Pageant, which did not ridicule the contestants but rather "prob[ed] the organizers to reveal how sexism is perpetuated

1.19

Optic Nerve. Left to right: Lynn Adler, Michael Cousins, Sherrie Rabinowitz, John Rogers, Mya Shone, Jules Backus, c.1975. Courtesy Lynn Adler.

in society."[112] This description is significant with regard to the politics of technology that would underlie Mobile Image, for it emphasizes the *processes* of embodiment and mediation. A feminist technology had to present the perspectives and experiences of women while also critically engaging the patriarchal logic of identity formation and its representation, the perpetuation of difference through categorical identification.[113]

Invited by the Bay Area Video Coalition to reflect on "Latinos and Media Art," artist and writer Coco Fusco cautioned against the commodification of cultural identity and "artworlding" of social subjecthood: "Despite frequent references to race in the alternative arts sector, the term should not be taken as a formal category in itself, nor as an ontological condition. The experience of race is relative to who and where we are, to other categories of experience we may inhabit, such as gender and class, to histories of differentiation, prevailing attitudes and political exigencies. That it is most frequently presented as a 'minority' issue is above all indicative of the desire to make it one."[114] The technological should be understood as a tool less for the representation of an experience than for the construction of one in all of its multiplicity and complexity. In her discussion of this construct*ing*, Fusco cites Stuart Hall: "Cultural identity . . . is a matter of 'becoming' as well as of 'being.' It belongs to the future as much as to the past. It is not something which already exists, transcending place, time, history, and culture. Cultural identities come from somewhere, have histories. But, like everything which is historical, they undergo constant transformation."[115]

Despite the joy and energy of collectively working on efforts such as Ant Farm's by now legendary 1975 *Media Burn*, Rabinowitz soon felt the aforementioned limits of and disappointment with video and guerilla television production. As Youngblood put it, recounting a conversation with the artist: "It had become clear to her that 'the medium' wasn't video, it was the image environment. Television was an image environment, and that's how one had to approach it . . . Continuing to produce tapes no longer made sense to her, because it did not affect the environmental quality of the broadcast."[116] Mobile Image's concern with the limits of broadcasting and their collaborative turn toward the aesthetics and mechanisms of mediation resonates in media theorist Alexander Galloway's recent differentiation between media and mediation. In his 2012 book, *The Interface Effect*, Galloway argues that the much of the historicization and theorization of new media has been hampered by a "conservative" notion of technology as "hypomnesis," of media as the "externalization of man into objects," of *techne* as "substrate and only substrate."[117] This history thus posits new media (in Galloway's take, the computer) within a dichotomous genealogy of the representation of the real and the reification of the technological, of technology as object and/or creator of objects. Instead, he argues, we need a theory of media that frames *techne* as process, as "lived practice." Mobile Image can be understood as having done just that in their investigation of the televisual as an operational logic, as neither solely an image of the world nor the world itself but, as Galloway puts it, "an active threshold mediating between two states."[118] Accordingly, to Mobile Image, it would not suffice to "arm everyone with portapaks" but rather to explore and engage with "the environmental nature of television, its power as an environmental process rather than discrete product."[119]

As a practice, *Satellite Arts* struggled with many of the questions driving contemporary media discourse today, while its (art) historical framing is, in turn, determined by them. As new media scholars Wendy Hui Kyong Chun and Lisa Nakamura have argued, "to read the digital in the non-digital" asks us to rethink media not so much as a more or less successful or faithful correspondence/mediation of material and immaterial realities, of private experience and public discourse, of bodies and images, but rather as a practice and politics of mediation *as* reality.[120] This view has profound consequences for the interaction of subjects as and with subjects and objects, and the formation of subjecthood as agency in that reality, because it posits telecommunications at the heart of such a practice and endows its artistic re-functioning with great power and responsibility. The implications are profound, suggesting, as Chun argues, that "the democratic possibilities for [new media] lie less in the creation of an ideal public sphere in which everyone is empowered and secure, but rather in its possibilities for messy and leaky engagement with others."[121] Chun's point recalls Mobile Image's rejection of "the democratic mythology of public access television," of "humanistic sentiment or good intentions." As a process and practice of and on reality, mediation was always already an engagement with others but, more often than not, governed by clearly (and cleanly) outlined demarcations of the material and the immaterial, experiences and images, bodies and languages and, consequently, subjects and objects, selves and others.

Interaction is at the heart of the "image as place," as is *embodiment*. Information and computer scientist Paul Dourish explains the relevance of embodiment to the understanding of televisual technologies: "By embodiment, I do not simply mean physical reality, although that is often one way in which it appears. Embodiment, instead, denotes a form of participative status. Embodiment is about the fact that things are embedded in the world, and the ways in which their reality depends on being embedded."[122] Any rejection of the myth of the public sphere as technologic-democratic ideal (and of the technological space as public) must therefore critically engage the *quality* of (dis)embodied interaction: How are subjects and objects embedded, and how do they interact? According to Nakamura, the history of telecommunication technologies, culminating with the Internet, has to contend with the neoliberal desire for disembodiment in the virtual and the depoliticization of the disembodied eye in the material.

The history of new media, explains Nakamura, is marked by the Internet's mid-1990s turn into a mass medium and an active politics of deracialization. The New Democratic platform of what would become the Clinton–Gore administration was based on "a neoliberal discourse of color blindness," a "universalizing discourse," which, especially in the wake of the 1992 Los Angeles riots, "proved extremely popular as it allowed avoidance of all discussion of race in favor of concerns that were perceived as more 'universalist.'"[123] Extending the utopian new media ethos of the 1960s and the marketing of the personal computer as a tool of both individual liberation and participation in a democratic process of cultural production, the Internet was framed around core principles of "privacy, competition, lack of regulation, and 'nondiscrimination,'" thus actively suppressing and perpetuating inequalities regarding access, subject positions, social spaces, and power relations based on race, gender, and class.[124]

Looking back at Mobile Image's underappreciated yet prescient attempts to investigate the possibilities and limits of embodied image environments critically, the question becomes: How did their re-functioning of existing telecommunication technologies affect the production of racial and gender identity formation? The extensive notes that accompany the planning and execution of *Satellite Arts* do indeed discuss the project in universal terms. Despite showing an acute awareness of the discrepancies between public access to and private ownership of communication tools—along with posing incisive questions about what constitutes democratic access, the power of imaging technologies as social organizing forces, the desire for "flexibility and dexterity in the social infrastructure," and the need to provide transparency with regard to the complexities of its organizing principles and processes—Galloway and Rabinowitz never explicitly differentiate what they meant by "audiences," "users," and "human effectiveness."[125] Arguably, however, it is more important to consider how the work *functioned* in relation to questions of power, access, and identity, and how it is historicized accordingly. In the case of *Satellite Arts*, Mobile Image consisted of Galloway and Rabinowitz, the members of the Mobilus dance troupe, and the engineers and crew working on technical and logistical components of the project. The list of "participants" (figure 1.20) includes more than thirty people and nine collectives, grant organizations, and other institutions. Of the technicians, most appear to have been white, male, and skilled, reflecting the (ongoing) demographics of so-called high-tech labor, while the dancers made up the most diverse part of the group, with two Asian women (Kimura and Mitsueda), one white man (Hoffman), and one Black man (Stinson).[126] There were differences in access from the outset: while Galloway and Rabinowitz were the ones leading the endeavor and able to secure financial and technical support from outside sources, including NASA, the Corporation for Public Broadcasting, and the National Endowment for the Arts—sources certainly less available to artists of color at the time—the dancers were both subjects and objects of the performance, while the engineers and crew operated the material apparatuses of image making, manipulation, and transmission. This is, in part, how things (including images) are embedded in the world; indeed, Lisa Parks defines visual capitalism as "a system of social differentiation based on users'/viewers' relative access to technologies of global media."[127] Since Mobile Image was dedicated to turning broadcasting technologies into tools of communication, *Satellite Arts* has to be examined as a network of relative agency, both of the participants toward the apparatus and of the participants toward one another: Who and what is the subject of interactivity; what is its object, now that more

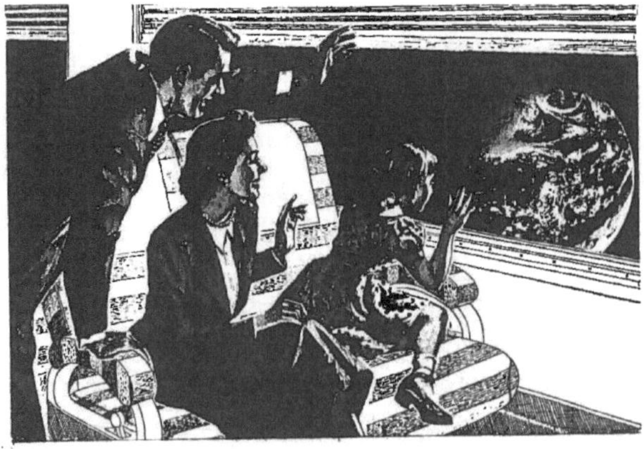

MOBILE IMAGE SATELLITE ARTS PROJECT

PARTICIPANTS

KIT GALLOWAY
SHERRIE RABINOWITZ
RICHARD LOWENBERG
JIM WISEMAN
BILL HEARN
MOBILUS DANCE TROUPE
  SOTO
  KEIJA
  MITSU
  NATHAN
PAUL HORN
DAN STAT
STEVE CHRISTIANSEN
TOM TUCKER
JAMES NIXON
NICK BERTONI
GERRY WURTZBURG
BILL HOWELL
DAN SORBI
DAVID GREEN
JOHN CHITWOOD
BRAD GIBBS
MYA SHONE
CURTIS SHREIER
LYNN ADLER
DAVID ROSENBOOM
JIM BARNES
JOHN PRESTON
MICHAEL COUZENS

BAIRD BROWN
SKY GARNER
KAREN MC LELLAN
ONE PASS VIDEO
OPTIC NERVE
CAL VIDEO
EDUCATIONAL TV CENTER
SOUNDS REASONABLE
STANFORD CHILDRENS HOSPITAL
NATIONAL ENDOWMENT FOR THE ARTS
CORPORATION FOR PUBLIC BROADCASTING
NATIONAL AERONAUTICS & SPACE ADMIN (NASA

And thanks to:
Public Interest Satellite Assoc. (PISA)
Andy Horowitz
Burt Cowlan
Wasyl Lew
Howard Wise/Electronic Arts Intermix
Cable Arts Foundation
Bob Kanter
Bill Etra
Skip Sweeny
Bertil Petersson
Stanley Greig
Allan Rabinowitz

1741 VIRGINIA STREET, BERKELEY, CA. 94703 TEL: (415)843-6335

1.20

Mobile Image, participant list for *Satellite Arts (The Image as Place)*, 1977. Courtesy Rabinowitz/Galloway Archive.

traditional positions and dynamics of viewer and viewed (of who is the "subject" and "object" of representation) are supposedly upended? For Galloway and Rabinowitz, there was to be, by design, "no disembodied eye," that body-less viewer and viewing position that underlay the intra-aesthetic discourses and practices of modernist visuality. If there was a *Lehrstück* quality to the performers' participation, a precursor to the Internet's "interactivity," then, according to Nakamura, "rather than focusing on the idea that women and minorities need to get online" or, in Mobile Image's case, in front of and behind the camera, "we might ask: How do they use their digital visual capital? . . . It does not change everything, but what does it change?"[128] What is the relative access to the organization of existing and possible experiences provided by Mobile Image's project?

From a technical position, the dancers remained subject *to* the apparatus, as others provided the means and knowledge of partaking in the mediated encounters. From a technological perspective, they are both subjects *and* objects, of and in the dynamics of mediation and experience. It is their bodies that act and react within the constellations of feeds, encounters, and environments. Within the given parameters and frames, there is the power to respond, to engage and disengage, testing and making visible the technical-material and aesthetic limits of the assembly. The organization and reorganization of experience as racial and gendered formation is not predicated on some mythically all-empowered subject (individual), or immobilized object (thing), and nor does it happen in a space that either claims or functions as neutralizing, universal environment. While the dancers remain subjected to the gazes of Rabinowitz and Galloway and the engineers and crew, they see themselves seeing and seen, on screen, next to it, among the cables and green screens, on the monitor, and so on. It is, in a sense, a double or split gaze. To that extent, it constituted what Nakamura calls a "volitional mobility" through different identity positions—positions determined by one's place behind or in front of the camera, at the console, at the drawing board—the boundaries of which cannot be easily (symbolically) transcended. Rather, they are comprehended and thus materialized as a site of struggle over access and power.[129]

Satellite Arts revealed the conditions of technological access as *relative*. A *Washington Post* article covering the project at the time ends with a quotation by Stinson while "standing among NASA's dish antennae": "They're not going to send me to the moon."[130] Although it is unclear in what context those words were uttered, they do underscore the confines of said mobility with regard to Black bodies in space at the time.[131] The mobile, double gaze in *Satellite Arts* worked to combat the "myth of interactivity," the experience of the imaged, disembodied self as liberated into and through the boundless ether of public sphere. To recognize oneself as a Harawayian cyborg, as mediated and mediate-able, is the first step to claiming (ownership of) selfhood in interactivity, to, as Nakamura puts it, "negotiate . . . identities as digital objects and in incremental ways move them toward digital subjecthood."[132]

If, in a world of "inhabited by wired, technologized, privileged subjects," the fantastic goal of technological innovation is, as Chela Sandoval puts it, "technologies developed by subjugated populations to negotiate this realm of shifting meanings," then the early satellite experiments of Mobile Image laid crucial groundwork in drawing awareness to the range of movement, of bodies and eyes, in and through interactive, televisual space.[133] Those experiments also showed that to "develop" technologies is not restricted to the production technical advancements, but must include the utility of existing apparatuses and tools according to their apparent as well as latent possibilities. The racialized and gendered bodies of the dancers were superimposed over a live view of the control room, over live television feeds, over an open field. The dancers saw themselves as part of a larger arrangement of images and bodies, tools and parts, the technical and technological constituents and agents of the mediated environment. The work neither merely reinforced given boundaries of access along lines of race or gender nor symbolically erased them. In some sequences of the work, split screens dissolved, merging the environments. Yet, the superimposed bodies and images of the dancers never quite fuse into a sort of universal hybrid being. *Satellite Arts* did not provide a deracialized ideal of (dis)embodied wholeness. Side-by-side and eventually overlaid, faces, arms, and torsos formed leaky constructs of multiple selfs and selves (figure 1.21). As a conscious process of bodily engagement in space, the work showed that, to quote Nakamura once more, "object and subject are not mutually exclusive roles: it is not possible to definitely decide who is being *interacted* and who is being *interactive* except

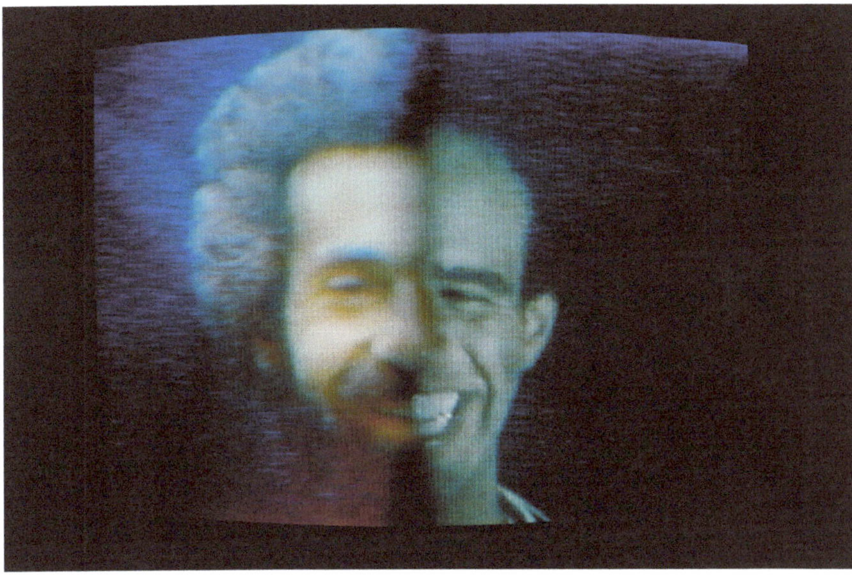

in specific circumstances."[134] *Satellite Arts* functioned to provide some of these circumstances, of situated knowledge, or the knowledge of being situated—of discerning not between bodies and bodies, between bodies and images, but between given and possible composites and processes of self-construction. In a sense, it forged a *re*embodiment in Dourish's sense: a making visible and tangible the embeddedness of selves and bodies, in front of and behind the cameras, before, during, and after the work—the participants' agency and lack thereof, as subjects and objects in the image as place, as agents of and subjects to movement in the image environment, as purveyors of technical know-how and capital, of access to the apparatuses of mediation.

1.21

Mobile Image, *Satellite Arts (The Image as Place)*, composite face, 1977. Courtesy Rabinowitz/Galloway Archive.

INTRA- AND INTERSUBJECTIVITY

Central to the innovative capacity (in a Brechtian sense) of *Satellite Arts* was not only that an analytic and productive relationality was established between ostensibly separate categories of making and viewing, subject and object, self and world, but also that this newly palpable "perceptual continuum," to use Miriam Hansen's term, this flow between traditionally dispersed places and positions, would drastically alter the very concept of selfhood and, with it, the potential for radically new social formations. As a TV of attractions, the work provided an experience of the apparatus and its spaces of production and reception as social, negotiated, interactive, and territorial. Hansen expands Gunning's "attractions" model: more than just a self-conscious experience of the media arrangement, it is a "site of overlapping, uneven, and competing types of publicity," a combination of both the technologically mediated and the face-to-face, generating an alternate horizon of social attitudes and actions and at least the possibility for "greater awareness of exhibition and cultural intertexts."[135] In *Satellite Arts*, panoramic forms of publicity were intertwined with public and private experiences predicated on relations that were embodied, immediate, and intimate, communicative and interdependent. Conventionally discrete sites and activities—watching broadcast TV, strolling in the park, dancing, virtual touching, operating in a control room, standing in front of the camera, standing behind the camera—were playfully, yet insightfully, connected in their determinacy. Like the early cinema, this conception of satellite TV as communicative, in-time interaction created, in Hansen's words, "a margin of *improvisation, interpretation, and unpredictability* that made it a public event in the emphatic sense and a collective horizon in which industrially processed experience could be reappropriated by the experiencing subjects."[136]

The attempt to accentuate and organize sites of reception as theatrical, performative spheres of production was by no means restricted to the work of Mobile Image at the time. In the early 1970s, before meeting Rabinowitz, Galloway was part of the Videoheads collective, founded by Jack Moore (figure 1.22). In 1971, Galloway, Moore, and other artists and media activists from Holland, Germany, India, Turkey, and the United States came together to build a multimedia theater and video laboratory

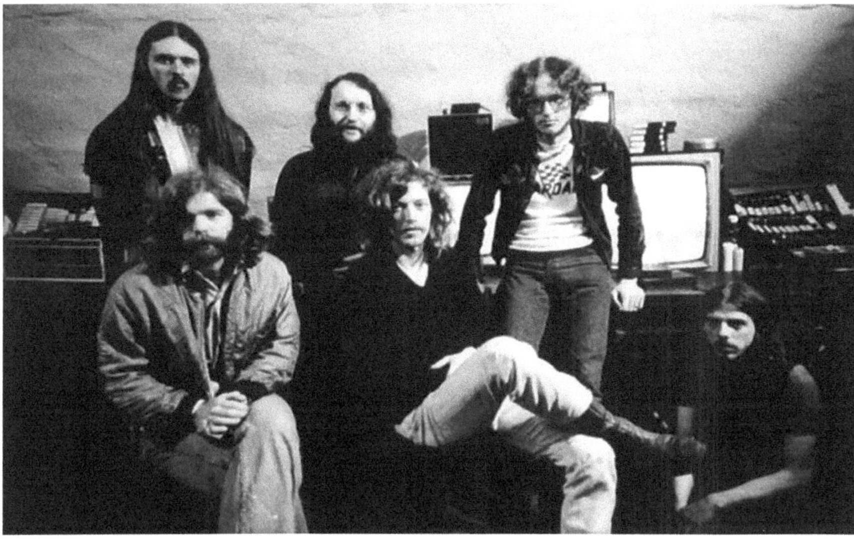

in an abandoned milk factory behind the National Ballet Theater in Amsterdam. Called the *Melkweg* (Milky Way), it defined itself as a "biotope of a variety of subcultures," actively opposing film and the broadcast media's "division of film makers, technicians, and the public."[137] *Melkweg* became a center of alternative video work in Europe. Comprising live video presentations, multiple monitors and projections, and transmissions between different rooms in the sprawling facility, its shows emphasized the process of media making and viewing as collaborative, social efforts, with feedback loops functioning as memory, reflexivity, and development. As Galloway recalls: "We wired the building with video and documented the events in it. Five nights a week I gave video shows using machines of different national standards on multiple monitors. It was also a cinema.

1.22

Videoheads collective. Left to right: Dan Foster, Terry Doherty, Jack Moore, Kit Galloway, Dave Jones, and Kirke Wilson, 1972. Courtesy Rabinowitz/Galloway Archive.

We built a modular stage for theater, wrote theater pieces, live video presentations. It was an entire media culture . . . we were already in the communications aspect of it."[138]

In the United States, organizations such as the People's Video Theater (PVT) and EZTV similarly mined the productive capacities of collective viewing. Founded in 1970 in New York by Elliot Glass and Ken Marsh, PVT was a video journalism collective focused on the social possibilities of the medium. Its members used Portapaks to produce gritty "street tapes" meant to capture the views of typically marginalized people asserting a public presence, from dialogues among and between different ethnic groups to mass demonstrations—the first Women's Liberation March, the first Gay Pride March, a Puerto Rican liberation group's occupation of a Manhattan church, an action taken by Native Americans at Plymouth Rock on the 350th anniversary of the Pilgrims' landing.[139] Those appearing in the videos would be invited back to the group's loft (its "theater") to view and talk about the tapes, with these conversations, in turn, recorded and replayed at other, often thematically related events. PVT transformed the place in front of the television set from a site of consumption to one of animated communication among various actors and participants in the process, and between several levels, of production. In 1979, artist and screenwriter John Dorr founded EZTV, a Los Angeles–based "video theater" that likewise facilitated collective viewing among the marginalized residents of its community—in this case, the historically gay neighborhood of West Hollywood. As Julia Bryan-Wilson has shown, EZTV created a flexible viewing space, incorporating malleable arrangements of screens and spectators to establish a "specific spatiality" that was "not only geographic but social—an 'out' gathering that included sharing skills, creating alternative representational practices, building communities, and fostering desire." Crucial to this new, queer(ing of) technology of productive-distributive sensibilities was the strategic upending of a range of reified viewing conventions—"formal" cinema-type screening, private television watching, avant-garde experiments in immersive or phenomenological film, and norms of gallery and museum exhibition, according to which works of video art are either navigated like fixed sculptural objects or integrated into display arrangements.[140] Dorr sought a new "spectatorial dynamics," as Bryan-Wilson explains, by combining and alternating theatrical, consciously staged modes of media presentation, including communal screenings, arrangements in which multiple monitors were to be moved around seated viewers, and attractions-type events that encouraged sustained attention without cinematic immersion.[141] Such models, Bryan-Wilson concludes, "meant to operate as a queer mechanism of incorporation, forming and re-forming tentative and temporary publics."[142]

The type of theatricality epitomized by these endeavors—in which improvised, interpretive, and unpredictable encounters were designed to produce a subject recognized and performed as a technological self—was emphatically different from that of both the "literalist art" Michael Fried (in)famously took issue with in 1967 and the "narcissistic" video practices Rosalind Krauss criticized a decade later. To Fried, the minimalism of artists such as Donald Judd and Robert Morris provides a depoliticized, timeless, "biomorphic" assertion of self that does not lead to a critical awareness of the viewer as historical subject.[143] Krauss takes "artists' video" to task for failing to transcend conventional notions of aesthetic encounter and thus serving the bourgeois, panoramic appropriation of the outside world as affirmation of the whole, coherent self. She differentiates between the narcissism of "reflection" and the criticality of "reflexiveness": "Reflection . . . is a move toward an external symmetry; while reflexiveness is a strategy to achieve a radical *a*symmetry, from within."[144]

It is precisely this "radical," productive asymmetry, this reflexive dissociation, this spatial and temporal incongruity that *Satellite Arts* modeled as it layered and delayed, split and united, the body and bodies in the televisual-*cum*-communicational, where perceiving the self and the other would break open and perform new subject constellations.[145] At the individual level, each dancer was required to operate as a kind of split personality—conscious of his or her existence in, and continuous need to negotiate, multiple places, temporalities, and roles. Each had to maintain enough distance to be able to control his or her surrogate in the screen, to move appropriately in its space and in relation to the others occupying it. Each had to remain cognizant of his or her physical position relative to both the camera and the monitor, while the time delays required a continuous internal dialogue—and division of attention—between physical form and projected form, actual body and its televised version. This was the quintessential challenge presented by the work. As Stinson put it, "I was never really sure when that little fucker on the TV who was me was gonna do what he was gonna do."[146] Again, one could not lose oneself in the image as the panoramic viewer routinely did, leaving home and body behind to travel across space and time. As participatory as *Satellite Arts* was, by design, it could not achieve the degree of absorption—of disembodied mobility—experienced while watching an ordinary television show. Instead, the dancers had to establish a self-conscious relationship with their own images, occupying not only two places and temporalities at once, but also two bodies, two perspectives, two subjectivities. As Galloway and Rabinowitz recalled: "Our brains were forced to operate in different modes. It was an intricate psychological and perceptual model that demanded a new way of relating to communication, contact and self. Our images acted as stand-ins representing us in a space we could not fully enter."[147]

This psycho-perceptual effect, this experience of a multiplied body and splintered self, further distinguished the work from other satellite-based projects of the time. It is also why the complexities and limitations of the technology, particularly the delays, were so central—again, in ways that were very different from seemingly like-minded telecommunications work. *Send/Receive*, for instance, similarly incorporated and even foregrounded the obstacles it encountered but as compensation for a technology that was still in its infancy and, more emphatically, to bolster its polemic against the suppression of bidirectional television by those maintaining control over that technology. The implication was that should these obstacles be overcome, and should control ever be decentralized and public access expanded, people would be free to express themselves directly and to connect and collaborate with others seamlessly. The whole, autonomous subject was awaiting emancipation.

In *Satellite Arts*, technological limitations were theatricalized and performed—not as hurdles on the way to a utopian third space in the screen but rather as opportunities to recognize that subjectivity is always historical, mediated, and split, the self always captured by the technology it takes up. This was explicitly staged on the third day, when a construction crew accidentally severed the cable powering the Goddard Space Flight Center in Maryland. With no East Coast location to connect to, the West Coast group experimented with bouncing its own content off the satellite,

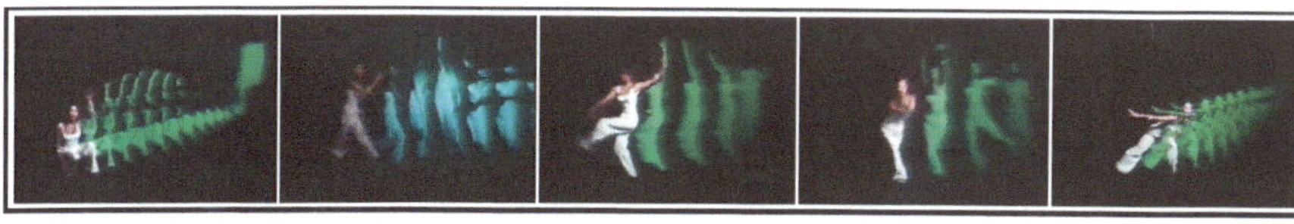

resulting in a series of scores in which single individuals interacted with themselves in the screen. In one, Kimura performed alongside a trail of ghost images, "dancing and choreographing an ensemble of dancers all of which are herself, each self-representing 1/4 second of her past" (figure 1.23).[148] She also wore wireless biotelemetry devices that monitored the electrical impulses of her body movements, which were automatically fed into an audio synthesizer that translated the movements into sound and sent them echoing through the satellite feedback loop. In this way, Kimura danced with multiples of herself, while making "live body music" from the motions of each of those selves. Toward the end of this score, the camera operators, video effects, and video mixer operators joined in, imposing their own improvisations upon her choreography. Kimura's body was re-presented as a container of proliferating selves, shaped by her own actions and those of others in the system.

The technological restraints of *Satellite Arts* thus opened possibilities for aesthetic experimentation and, more profoundly, for reflection upon the splintered self, its relationship to "reality," its position within the technological arrangement, and its negotiation of that position—the processes by which it mediates and is mediated. Such reflection carried over into the real lives of the participants. As Stinson put it in the immediate aftermath of *Satellite Arts*, "I noticed that separations are much more obvious in my life. And that I feel separated a lot. I'm aware of being separated a lot. I sit in my house and have feelings I haven't had for years like

1.23

Mobile Image, *Satellite Arts (The Image as Place)*, video stills, 1977. Courtesy Rabinowitz/Galloway Archive.

loneliness."[149] A few months later, in a letter to Galloway and Rabinowitz, he again mentioned the residual out-of-body effects of the experience: "I am still in time delay, still looking back. . . . at an image of my body flung out into space on the wings of a distant bird. At the memory of a man trapped in the past, one-half second removed from the moment. And it's difficult to move with my head turned over my shoulder."[150] The experience of coordinating movements, maintaining multiple foci, inhabiting multiple places, and occupying multiple positions within the apparatus underscored the fact that technology is both internalized, projected into the body, and externalized, as the body projects itself out by way of it.

Comprehended as a historical, technological self, the body in *Satellite Arts* functioned less as a timeless, total subject–object, a fixed form, than as an instrument of orientation, negotiating between interior and exterior space. Indeed, as Negt and Kluge argue, the body is that through which all experiences—of both self and world, inseparably bound and mutually constitutive—are filtered, processed, and managed.[151] At once personal and social, the body is itself a tool, a technology, in that it functions as mediator and manipulator, even as it itself is mediated and manipulated. In *Satellite Arts*, the body is experienced in just this way via the work's moments of performativity and theatricality, coordination and communication breakdown, improvisation, and shifting relations, through its playful simultaneity of presence and absence, intimacy and distance. As Leslie Adelson explains in his discussion of Negt and Kluge, "things," including selves, "can never be embodied *per se*, since they always take the form of socio-historical, that is to say subjective–objective *processes*."[152] Consisting almost entirely of such things and selves in and as process, *Satellite Arts* demanded a conception of multiple, often antagonistic subjectivities within the same body, rather than the coherent, individuated self and body as posited by televisual logic, panoramic sensibility, the idealized and projected "inner sovereign" of bourgeois society. In the truly telecommunicative experience, the subject—indeed, identity as self and other in the world—is constituted as and by an *Eigenschaftskette*, a chain of shared, obtained, and missing attributes, relative but not relativist, as they are specific to particular, historical arrangements of time and space.[153]

To effect this experience, participants in *Satellite Arts* were routinely compelled to step in and out of their conventionally prescribed spaces and roles. In "Triangle Dance," for example, Stinson did not appear in the screen but instead interacted with the three dancers from an unseen position outside of it. Taking advantage of the delay, he intervened in the composite visual space by "playing" the audio feedback loop like an instrument. The dancers responded to his ethereal soundtrack—whooshes, clanks, rattles, chimes, hums, haunting moans—and vice versa. At times, Stinson was able synchronize the sound with the activities happening in the screen; he inserted a bang, for instance, when Hoffman swiped his hand at Mitsueda, and again when Mitsueda kicked back. At other times, the dancers shouted out, interrupting his soundtrack. Later in the score, Stinson attempted to produce more harmonic sounds by chanting progressions of operatic tones and attempting to sing along with the proliferating layers of echoes and echoes of echoes. Overall, he served as a kind of external God-like force, but one whose own interventions were, in turn, captured by the system and subjected to its alterations. The "composite" thus comprised not only the two televised spaces but also the on-screen and the off-screen, reinforcing the fact that the latter, and those inhabiting it, were very much part of the score. In this sense, all participants—the dancers, seen and unseen; the camera operators spontaneously zooming in and out; and the engineers balancing the audio feedback—were operating as an improvisational performance troupe, at once contained by the apparatus and free to move within and cut across its various dimensions. This dynamic pervaded the entire project, from the control room relaying live broadcast imagery into the screen to the performers using the setup to talk back to and influence the production team. In an "evaluation/debriefing session" soon after the completion of *Satellite Arts*, Rabinowitz and Kimura recognized the effects of this on how participants came to understand the arrangement, how they became conscious of—and began to defy—the conventional distribution of control:

SHERRIE: One of the best things for me was when Keija yelled out during Time Delay Feedback "get that thing out of there I don't like it," and it was taken out.

KEIJA: I was feeling my power on the last day.

SHERRIE: But that was good. For me that was a success, that you felt that you could control the control studio.[154]

This sense of mobility and control opened up possibilities to contemplate the televisual relegation of particular bodies and identities to particular functions in particular spaces. Dressed in generic costumes of white long-sleeved shirts and white pants, the performers in *Satellite Arts* appeared to interact with images, environments, and one another—and from multiple positions of authority and intimacy—without regard for conventions of gender, race, and ethnicity. Yet, this sense of sameness and interchangeability remained only partial. Dancers were variously superimposed and separated, melded together and distinguished from one another, collected and contrasted; dissimilarities emerged even as distinctions were blurred. This happened throughout the work, and it was a determined component of the early rehearsals, as some of the first experiments with a composite screen involved dancers attempting (and failing) to match up with each other's bodily contours. Just as the experience of being simultaneously in two places and two times, or in multiple roles within the technological arrangement, caused participants to consider the mediation of space and time and position, the experience of being at once the same and different generated reflection on categories of identity, of difference, and fantasies of likeness, of commonality. The dance troupe was rooted in the San Francisco Dancers' Workshop, where, in line with the philosophy of its founder Anna Halprin, performers were trained to examine questions of authority and difference, place and context, in their process-oriented performances. Like Mobile Image, Halprin considered her workshop to be a place of experimentation, where movement through space was intimately bound up with social terrain, and dance was understood as an investigation into community and the individual and collective rituals that sustain it. Galloway and Rabinowitz chose Mobilus partially because of this approach, and the very structure of *Satellite Arts*—its sequence of task-oriented "scores" set within a workshop format—was borrowed directly from Halprin. "Scores," as she explains, "allow for individual input within an ordered collective whole . . . The purpose is to create change . . . to find out what our differences are and what our commonalities are."[155] Again, the work did not simply seek an elision of these differences. Instead, it sought to both stress and transcend markers of identity, to present them as constructed and enacted, both ideological and real in their materialization and rituality, and hence establish a site where regular rules at once applied and could potentially be restructured. This was the essential lesson in the communicative distribution of authority provided by *Satellite Arts*: to realize how much and how little power one has and whether given arrangements could be altered through action and interaction, through communication, as opposed to the performance of conventional attitudes, perspectives, and behaviors. Dance—and, by Galloway and Rabinowitz's extension, every activity involved in their project—became the enactment of roles and their potential transcendence. To Halprin, this method was seen as decidedly sociopolitical, applicable to interpersonal relations and to a range of issues, from conflicts over race, gender, and sexuality to city planning, community formation, and environmentalism.[156] As experienced by the participants of *Satellite Arts*, the kind of mobilization and critical reflexivity enabled by the work facilitated a conception of identity as performative, making visible the televisual separation and mediation of the raced and gendered bodies interacting in the screen.

The experience and conception of a constructed, malleable, shifting intra-subjectivity produced by the "perceptual continuum" of *Satellite Arts* suggested possibilities for a different notion of *inter*subjectivity. From the start, the project was understood as an experiment in a new sociality, theatricalized in ways that generated critical consciousness of hierarchies and organizing principles. As Galloway and Rabinowitz recalled, "One of the most rewarding aspects of the project was the process of putting it together and the creative interaction that took place not just between performers or performer-director but between performers and engineers, NASA and ETVC administration and artists. The nature of the project demanded a new social structure and integrations of usually separate disciplines/realities. Everyone came away from the project feeling we had just begun and wanted to continue."[157] Although they may not have used these terms at the time, Mobile Image thus recuperated an avant-garde history of the artist engineer in the Constructivist and Bauhaus sense; their collaboration with industry and government apparatuses sought to transcend the simplistic binary of collusion versus subversion and critically analyze methods and logics of production, distribution, and commodification in order to potentially transform them. This "new social structure" extended well beyond the transgression of defined jobs and disciplines to the interpersonal communications that took place in the screen. As a gathering of subjects made to recognize and perform as divided, technological selves, the social field established by *Satellite Arts* consisted of a constellation of splintered subjectivities, theatricalized as such. While they grappled with being in multiple places, times, and positions at once, participants were also abundantly aware that those with whom they sought to connect were themselves grappling with the same situation. As Galloway and Rabinowitz put it, "The performer was also a viewer of a program in which their disembodied image was performing with a disembodied partner."[158] Again, the objective was not to produce some fantastic display of seamless, technologically enabled collectivism but rather to stage the intra- and intersubjective complexities of communication and test out its effects, the results of which were decidedly not predetermined. "The connection seems tenuous," Galloway and Rabinowitz observed. "Does it bring them closer together or does it make the separation more apparent? The edges of definition blur."[159] Or, as Mitsueda put it, "I just got hit with a paradox of trying to communicate and create and creating distance through technology . . . the paradox of trying to communicate through technology and feeling the isolation and distance created because of trying to communicate through technology."[160] One score, "Monitor"—rehearsed and planned but not executed due to the technical difficulties on day 3—was designed to accentuate this paradox, the notion of communication as a self-reflexive struggle to operate in a social field in which everyone is both mediator and mediated. As Galloway and Rabinowitz described it: "The focus of this piece is the monitor as conduit for and container of the other person. A dancer at one location is shown holding a TV monitor, talking and reacting to it. The dancer coming across the monitor reacts to him. This piece most clearly demonstrates the phenomena and experience of communicating and dancing with a partner who is only there on TV."[161] Here, the television is materialized—not just as an object in space, as in works by Paik and others, but also as a "conduit for and container of" the communicating subjects. The score was to function as a kind of meta-examination of the very social situation being established by *Satellite Arts* overall.

Television was thus conceived and used as a tool of both connection and separation, and interaction through it took place among subjects who were each a tenuous and shifting amalgam of multiple selves—physical, technological, social—empowered and controlled, liberated and imprisoned, continuously framing and being framed, mediating and being mediated. The new social formations modeled by *Satellite Arts* were therefore made up of various and variable groups of individuals, each a "public" of overlapping, contending attitudes and attributes, extending numerous possible "chains of attributes," to borrow Negt and Kluge's term, through and from each subject to others. Such bodies and subjects as processes *assemble*, not just in the sense of individuals gathering together. Rather, each formation's capacity for an *emancipatory* shared or public subjectivity is constituted *in* social process, as the assembled subjects no longer resist but give in to the pleasures of recognizing relationality as something that is not only a concept, a consciousness, but also a physical and psychological reality.[162]

Judith Butler's recent treatise on the politics of assembly sheds light on both the possibilities and limits of *Satellite Arts* as a model of an emancipatory communicative practice. For Butler, the assembly of bodies, virtual as well as physical, is a political practice when the bodies affirm their plurality through performing a "being-with" in a process of becoming and making—meaning that such bodies are public, are formed, actively and passively, in process with other people, things, images, and so on. This performance becomes what Christoph Menke calls a "social *counter*-practice of the body" when it is being "theorized" *and* when it is inscribed in a theory of contemporary society.[163] Theory here refers to a theatricality, a performance of assembly that critically reflects unto itself, that mobilizes perception to such an extent that it becomes reflexive, an assessment of self in sociality, in space, in time. As James Clifford puts it, "To theorize, one leaves home."[164] In *Satellite Arts*, participants were compelled to travel beyond familiar locations, points of orientation—a journey of critical displacement unlike that provided by the panoramic, entrenched, fixed roving gaze of the conventional satellite broadcast eye (exemplified by *One World*). It is this displacement, this alienation through mobility that, in turn, sheds light on the precariousness of the privatized body, isolated, relegated to particular spheres of selfhood and otherness, marginalized, individuated as disempowerment, lack of solidarity, of connectivity.

Ultimately, however, *Satellite Arts* was an exercise more in assembl*ing* than in assembl*y*, its laboratory-like conditions predicated on as well as limited to "theorizing." The multiethnic, multi-gendered group of dancers connected through the mediation and mobilization of the work but did not attempt to establish a conditional relationality between each participant's existing and potential "chain of attributes," the specific experiences and circumstances of what Butler calls "precarity"—the dispossession of selfhood, of the production and distribution of subjectivity, individual and social. Crucially, Butler's term refers to the particular everyday violence disproportionately visited upon the "subjects" of neoliberal society. It is "that politically induced condition in which certain populations suffer from failing social and economic networks more than others, and become differently exposed to suffering, violence, and death."[165] Assembly becomes a social counter-practice when it asserts bodies as plurality in interdependent presence, and when it links to other assemblies and struggles. To become a communicative practice in this sense, the work of Mobile Image would have to establish a relationality between the performative politics of technology, as presented in the *Satellite Arts* laboratory, and the ways in which such performativity has played and could play out at specific moments, in specific places, among specific people. Only then would such work expand the struggle over perspectives, over the social horizon of experience that enables or cripples our capacity to relate, organize, and transform.

2

HOLE IN SPACE: ELECTRONIC PUBLIC SPHERE

In November 1980, Mobile Image unveiled *Hole in Space (A Public Communication Sculpture)* (figure 2.1). The work consisted of two large screens linked by a live satellite feed and viewable from sidewalks in front of Lincoln Center for the Performing Arts in New York City and the Broadway department store in the Century Square Shopping Center, an open-air shopping mall in Century City, Los Angeles. For three evenings, two hours per day, participants could see, hear, and interact with one another in real time from opposite sides of the country. Unannounced and with few preset rules, the work aimed to offer a social experience that demonstrated what the artists called a "globally distributed electronic commons."[1] People stopped and gathered, waved hello, played games, asked questions, sang songs, made new friends, and reconnected with old acquaintances. At first, the encounters were entirely by chance, followed by prearranged reunions over the subsequent days. Rather than set up a virtual location, an "image as place," as they did in *Satellite Arts*, Mobile Image dissolved nearly three thousand miles of space, creating, in their words, "an outrageous pedestrian intersection."[2] Although the interactions it enabled were at times superficial and fleeting, *Hole in Space* fostered critical considerations of broad communication mechanisms—their relation to public and private life and "public sphere" configurations, the apparatuses that constitute and are constituted by them—while making palpable and visible the limits and potentials of such mechanisms.

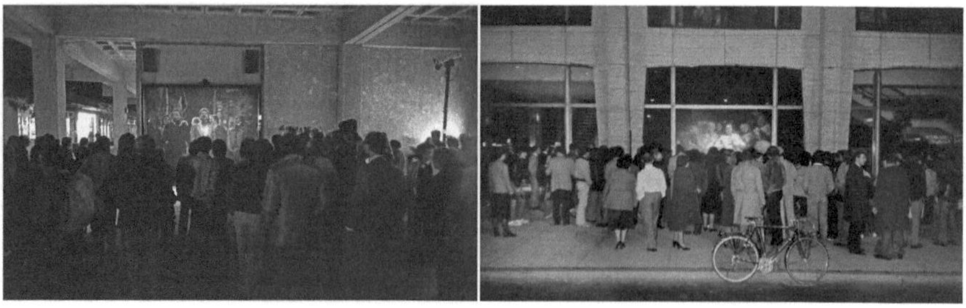

Mobile Image thus transitioned from experimental art and performance to "real life" intervention and interaction, from an arrangement accessible by select actors and operators to direct public use and the social possibilities generated by relatively unrestricted communication systems. Enabling participants to imagine an alternate historical present in which bidirectional television was commonplace, *Hole in Space* can be seen an extension of *Satellite Arts* into the everyday world. The work was similarly conceived as a laboratory in which participants could playfully test out reconfigured telecommunications technology—the television, along with other devices such as the telephone. As the artists explained, "We are concerned with the lack of human control and subsequent dehumanization that occurs when routine takes the place of imagination. Increasingly in this technological society, the public is led to believe that an application of a technology is an inherent and immutable characteristic of that technology. Software is mistaken for hardware—our public imagination and bank of apparent options is severely limited and those with most access to the tools have the unchallenged power to shape our image of the future. 'Hole-in-Space' is a separation of the technology from the habituation of its context."[3] Bodies interacted through the screen, resulting in intricate

2.1

Mobile Image, *Hole in Space (A Public Communication Sculpture)*, 1980. Courtesy Rabinowitz/Galloway Archive.

and overlapping improvisational social "dances" that, like the earlier project, required an internalization of the technology at hand, from its physical construction and boundaries to the inevitable time delay. Yet, *Hole in Space* was both simpler and more complex than *Satellite Arts*. Its setup was relatively static, its interface straightforward, its machinery mostly hidden from view, and it was grounded in the specificities of location, not only the two iconic American cities but also the particular urban spaces that were linked together. Because it was outside and ostensibly open to anyone, though, the work involved exponentially more variables—environmental factors, unexpected interruptions, chance encounters, and the chaos of multiple concurrent communications, activities, and audiences. Unrehearsed and unscripted, without any overt involvement by the artists and technicians who operated it, the project's direction—its form, etiquettes, and happenings at any given moment—developed on-site and through the social and situational dynamics of the crowd, activated by a sense of spontaneity and organic occurrence.

Mobile Image again underscored the generally "public" aspect of technology, establishing a "composite" space in Miriam Hansen's sense of the term—that is, involving both the material imbrication of distinct forms of publicity and the coexistence of multiple interactive and competing social and individual horizons of experience. With *Hole in Space*, however, they focused explicitly on the urban context and the ways in which it, public, art, and technology interrelate as entangled sites of struggle. The work coincided with a moment in the United States when ideas and images of the city, both as physical site and as arena of public life, encounter, and discourse, were highly contested. Advances in communications and information technologies were deeply embroiled in this contestation, as were increasingly popular practices of "public art." The late 1970s and early 1980s was a time of assessment and lament by certain critics and theorists over the so-called decline of the public sphere, while certain artists directly confronted both historical and then-current relationships between public and art. As high-tech devices and media were becoming more and more personalized and privatized, as people could increasingly sit in the comfort of their homes and access a world of information via cable television, satellites, video tape, and personal computing, technology continued to be seen as both fundamental and detrimental to social progress, isolating as well as democratizing. What had typically been associated with the public sphere—social connection, communication, chance encounters—could now supposedly be achieved without going outside, but that new capability came with preset rules and protocols largely determined and enforced by increasingly deregulated commercial interests. Meanwhile, new devices such as the Sony Walkman (released in 1979) enabled people to customize their public experience in ways that would become ever more prominent over the ensuing decades, with the proliferation of mobile telephony and other soon-to-be ubiquitous "technologies of mass individualization."[4] Public engagement was becoming something to be either purchased and received in the home or circumscribed by personalized media out in the world, the public sphere essentially being privatized both at the consumer level and in the increasing control over it by corporations, even as exciting new possibilities for greater access and participation were emerging. Whereas *Satellite Arts* addressed this conundrum in the abstract, *Hole in Space* homed in on the urban setting as the traditional site of the public sphere, as the place of converging material, ideological, and aesthetic forces that not only structure people's conceptions of communication, its potentials and limitations, but also impact the allocation and definition of space itself, of neighborhoods, districts, and the passages within and between them.

At first glance, *Hole in Space* can be understood as an attempt to reconstruct an idealized "bourgeois public sphere" through the democratizing function of telecommunications, and in fact it was likely experienced in that way, at least initially. Mobile Image presented a novel form of unstructured social interaction, involving both arbitrary and planned encounters, a town square for the New Media Age, reviving historical notions of urban public space by means of the very technologies that were supposedly usurping it. However, the nuances of the work—the particulars of its setup and chosen locations within the cities, its progressive unfolding over the three days, its configuration of participants and audiences, and the specific activities and behaviors it put on display—enabled it to engage public sphere ideals and the nostalgia for them critically, that mythical time and space in which, ostensibly, everyday people could equitably and effortlessly convene, discuss, debate, and function as a collective check on power. Although the response to the work was often euphoric, it also caused participants to reflect upon what was lacking amid the rising wave of technological possibility and choice and why certain potentials remained unfulfilled, as well as the reifying rhetoric surrounding so much new technology—and adopted by so much "public art." *Hole in Space* functioned as a complex, ambivalent, and ambiguous framing of publicity as a heterogeneous, shifting site of struggle over cultural meaning, the right to the city as an arena for the assertion and challenging of social relations, and the desire for mediation itself. Set within "shopwindow" frameworks into which pedestrians looked and through which they were seen, the work openly conflated the social via the screen with the commercial, suggesting that the mythological image of the bustling city with an intact public sphere was in fact an integral component of the modern, capitalist metropolis, where selling the fantasy of connectivity and participation serves to satisfy people's needs just enough to legitimize exploitations and exclusions—whether according to class, race, gender, or other categories of division.

At a moment when perennial questions about the relationship between technology, urban experience, and public life seemed particularly urgent, *Hole in Space* conjured up the long history and discourse of the public sphere. The work's image of futuristic telecommunications evoked yearnings for a more complete psychosocial totality associated with a supposedly better past, awaking people's almost childlike enthrallment with a brand-new toy. Yet, it did not simply provide what such a product ostensibly would: the gratification of symbolic liberation from the fragmentation of late-modern existence. Nor did it operate as a neo-avant-gardist negation of the emancipatory potential of artistic engagement. Rather, it functioned as a performance, a staging and enactment of commodifying mechanisms, of communicative behaviors—some conformist, others innovative—and thus of the ways in which certain arrangements of physical and discursive interaction and intimacy (the public) are capable or incapable of addressing and satisfying existing needs. Recognizing discrepancies between naturalized ideals and the particularities of time and place, participants could potentially begin to grasp the relationality necessary to transform "dream images" into "historical images," to borrow Walter Benjamin's terms, and thus envision new ways of understanding and ultimately organizing experience.

THE SETUP: TECHNICS OF ENCOUNTER

Hole in Space effectively operated on two levels. On the one hand, it functioned as a techno-sociological experiment, with the interconnected crowds as unwitting subjects. As Galloway put it, "I remember seeing a drawing of a big eye looking at a microscope. And on the glass there was a crowd of people. *Hole in Space* was like that. We were the eye."[5] The work thus continued Mobile Image's collaborative "research and development" practice, involving artists, engineers, high-tech devices, and users in an exploration of the possibilities and limits of telecommunication tools and the ways in which certain arrangements make visible perceptual norms, relations, and potential transformations at the moment of their enactment: "We wanted to see what people would do . . . We wanted to observe people having the experience of discovery. How would they receive it and acculturate it? What would they say and be and do? What would they invent? How would they define a social situation with no rules?"[6] To answer such questions, the system was designed to facilitate direct, frictionless social interaction. Whereas, in *Satellite Arts*, "transparency" meant deliberately exposing the work's internal mechanics—its constellations of spatial, technological, and human relations—here it meant something very different: people were to look through the high-tech arrangement. Constructed so that its intricate machinery would be as inconspicuous as possible, *Hole in Space* was decidedly polished, its visible equipment—nine-by-twelve-foot screens, fixed video cameras, microphones, and speakers—neatly contained within a storefront-style window. The rest of the work was concealed: there were no labels or announcements of its title or intentions; its "backstage" technicians, control rooms, and machinery (including refrigerator-sized rear projectors) were kept hidden; the artists themselves remained unacknowledged; even the videographers who documented the project were selected because they were experts in, as Rabinowitz put it, "being part of the crowd, using no lights, just letting the situation tell the story."[7] When people stood about twelve feet from the screen, they would appear life-size to their cross-country counterparts, and the arrangement was aligned so that each side would experience a semblance of eye contact with those they faced. To minimize distraction further, Mobile Image did not allot a small section of the screen in which participants could see themselves—already a convention of emergent teleconferencing systems—and they used infrared cameras and radiation emitters to illuminate the scene invisibly, eliminating the need for obtrusive artificial lighting. While the artists continued to insist that "the processes, communications, and relationships that make up the network are truly the armature of the work," its apparent simplicity and this lack of interference were essential to the concept of a technologically enabled "hole," an open channel through which spontaneous, unobstructed sociality could ostensibly occur.[8] However, whereas that sociality was seen as potentially transformative, it was presented not as some pure, unadulterated form of communication but rather as always already mediated.

Galloway and Rabinowitz hoped that the experience of *Hole in Space* would provoke the kind of critical awareness necessary "to liberate people's imaginations," as they put it, to inspire "an alternative to prepackaged and labeled fantasy" that would elicit consideration of why technology develops in certain ways and not others and how such development impacts the world, both physically and ideologically, locally and globally.[9] This was challenging because, unlike the Mobilus dancers who performed in *Satellite Arts*, participants in *Hole in Space* were random, untrained, constantly changing, and, generally speaking, not necessarily as predisposed to consider the work's sociopolitical and aesthetic implications. Indeed, they were largely unaware that they were taking part in an artwork at all—and the artists could assert very limited influence on participants' behavior: there could be no rehearsals, no pointed shifts in the broadcast, no preconceived "scores" to provide structure and focus attention. Yet, the selection of locations—not only the two American cities with their respective associations but also the particular sites within those cities—compensated, as did their management of the work's timetable, and their use of advanced publicity and solicitation of media coverage. The goal was to balance the apparent dissolution of space with an emphasis on geographical difference, simultaneously eliminating and accentuating place, time, and distance, and thus fostering consideration of the context in which the work was situated and the relationships between that context, advanced technology, and public life.

To heighten the disparities between the two cities, the artists chose to install their work in autumn, when weather contrasts would be obvious (but when the risk of a winter storm in New York was low), and they selected 8:00–10:00pm Eastern Time to emphasize the time and light differences, as it would be dark in New York from the start but would transition from day to night in LA during the two-hour span.[10] People picked up on this right away. Some of the first interactions were questions and answers about the weather and time. "It's thirty-eight degrees there," someone in LA declared. "You can tell by the way they're dressed!" Writer Pamela Lifton-Zoline, who chronicled her experience of *Hole in Space* in notes for an unpublished essay, recognized that the tendency of participants to begin by pointing out these basic differences reflected the paradoxical essence of the work: "Sometimes this was simply the most obvious orientation query for someone newly come to the screen, but the context indicated that these questions had more than that simple effect. They also served as orientation phrases in that they acted to position the interlocutors, to re-establish the sense of distance and attendant wonder in what was perceptually an immediate, distance-annihilating experience. They sharpened the central contradiction at the core of the piece, 'we're <u>here</u> and you're <u>there</u>, but we're together, face-to-face.'"[11] This was followed by jokes involving location-specific stereotyping and self-stereotyping, as well as clichés about such things as attitude, accent, and the supposedly adversarial relationship between the two cities:

NEW YORK: What're you doing in LA?

LOS ANGELES: Getting a tan.

. . .

NEW YORK: How many New Yorkers does it take to screw in a lightbulb?

LOS ANGELES: How many?

NEW YORK: None of your fucking business!

. . .

NEW YORK: Do you know the difference between Dannon yogurt and LA?

LOS ANGELES: What, what?

NEW YORK: Dannon yogurt has an active culture!

LOS ANGELES: The guy next to you just picked your pocket!

. . .

LOS ANGELES: You tell us everything you hate about us and we'll tell you everything we hate about you.

NEW YORK: There isn't enough time!

At the very moment of social connection, the two sides reflexively reemphasized their own real-world positions, along with the temporal, spatial, and cultural distances between them. For Mobile Image, this positioning was a crucial component of the work—that people did not just become absorbed into the screen or overly enthralled by the cross-country bonding experience but rather established, from the start, a self-conscious approach to what was happening. With *Hole in Space*, they explain, "you got to work both of those things out at the same time, the commonality and the difference."[12] "Do you realize we're making history?" a man in New York asked. "Do you realize we're standing here talking to a pane of glass?" responded a woman in LA.[13]

These initial "orientation phrases" also immediately established a version of the public based on local cultural identity as the basic social unit, one that initially seemed to subsume all other categories and which persisted despite the constant changes among those who constituted it. Participants were either New Yorkers or Angelinos, and, as evident in the initial rounds of jokes, they readily assumed the attendant mythos of those monikers. As will be discussed, the quality of these publics shifted over the course of the three days, but for much of the duration of the work—especially during the first two days—people instinctively took part as members of those groupings. And when spontaneous cross-country bonding occurred, it was likewise as complementary parts of a social formation. "At some point in the two-sided party," Lifton-Zoline explains, "the technology and the politics and the aesthetics came together . . . And the shape of the party was protean, like a gathering around the water hole, the market, the life raft, the bomb shelter."[14] Rabinowitz similarly recognized this dynamic: "They used their collective energy to create something that was more than their individual connections across the continent. They started communicating as a group rather than as separate individuals carrying on private conversations that no one else could relate to. Instead, they began relating together."[15] Even when people stepped forward from the crowd to express themselves or connect with another, they always, to some extent, did so also as ambassadors, as representatives—if not representations—of their location-specific group identities.

The social qualities of the at-first chance and subsequently planned encounters enabled by *Hole in Space* were not monolithic. Rather, they represented a collection of overlapping concerns, identities, and intentions, shifting effects and affects, inclusions and exclusions. On the one hand, these encounters evoked conventional bourgeois public assemblies; on the other, they provided opportunities for individual connection, for the establishment or reestablishment of private relations. Physical and electronic space was at once rendered public and presented as commercialized and the result of displacements. The crowds were emphatically local and emphatically placeless, at once enabled by technology and bound by it. At times, they coalesced; at others, they atomized or dissolved into chaos. Participants appeared both to crave authentic communicative experiences and to perform for the camera as media-constructed types, their interactions swinging between seemingly genuine communication and improvised spectacle. Conceived as a laboratory, as a social experiment under Mobile Image's "microscope," *Hole in Space* raised questions about what constitutes a range of core concepts—"public," "assembly," "gathering," "communication," among others—in the Media Age.

This was a vital part of what made the experience so unusual, of Mobile Image's innovative repurposing of technologies that were cutting edge, on the one hand, but not really so exotic, on the other. Large screens, live satellite transmissions, direct communications were already fairly common in 1980—whether in the form of nightly news broadcasts, teleconferencing systems, long-distance telephony, emerging computer networks, or giant stadium scoreboards, the first of which had just debuted that summer in LA's Dodger Stadium. What was extraordinary was not only that these technologies were combined and placed outside for free, open-ended use by anyone passing by or hearing about it and choosing to participate, but also that it raised questions about that very encounter, its parameters, and contextual relations. *Hole in Space* was conceived as a direct extension—and investigation—of the urban public experience rather than as a demonstration of bidirectional television or as a gigantic version of the videophone, that mythic, seemingly ever-anticipated advancement on the home telephone.[16] Participants had to stop and submit to a collective activity via technologies, most of which were typically relegated to (or imagined to be one day part of) the private sphere. Furthermore, the specific locations that were linked together—a cultural complex and a shopping mall—provoked consideration of how such institutions relate to telecommunications apparatuses and were themselves neither neighborhoods nor categorized working districts. Lincoln Center and the Century Square Shopping Center are places for leisure, to which people come for a particular event or need, or through which people simply pass. By situating *Hole in Space* in such places, Mobile Image enhanced the potential for arbitrary encounters and minimized the possibility that participants would already know each other, or that particular constituencies would assert authority over the space. These were plainly not the kinds of streets that Jane Jacobs celebrated in the early 1960s—places that organically bind communities together via networks of local shopkeepers, neighbors, and regular pedestrians, places that shape the "public identity" of people according to shared needs. Furthermore, the work's unannounced midweek schedule meant that people had to interrupt their routine moments of personal distraction—their commutes home from work, their shopping habits, their weekday evening leisure time—to join in.

The experience of *Hole in Space* was thus one of continuous adjustment and accommodation to a variety of variables: the space and urban context in which the work was embedded; the parameters of the system; and the people with whom one shared that space, context, and system. Unlike movies, television shows, or other spectacles through which people lose themselves—or the telephone, in which individuals are engrossed by the intimacy and privacy of the conversation—this assembly involved constant negotiation between absorption and self-awareness. Participants had to stay mindful of the fact that they were on view as much as they were there viewing, always at once performers and audience. Whereas they could not see themselves on the screen, they had to remain cognizant of their physical position within the arrangement and how they were being seen on the other side of the country, as well as their relationship with those standing beside them and the broader dynamics of their respective urban locales. This constellation—participant-technology-cohort-counterpart-city—was underscored by the rigidity of the work's setup, which required that people stand a certain distance from the window, know exactly where to look, and remain aware of the fixed boundaries between on-screen and off-screen space. Although the camera was visible and lines were drawn on the street at the twelve-foot mark, the arrangement was largely regulated by the crowds themselves, which determined who spoke when and continuously kept participants in check by admonishing them to move back, step forward, look down, or shift into view. This feedback loop also incorporated much more than physical orientation, as virtually every action elicited responses—laughter, encouragement, cheers, jeers, reprimands—from both sides, home and away (figure 2.2). A microculture of surveillance and self-surveillance arose, in which even nonaction was fair game, as passive observers were periodically pressed to come forward, identify themselves, and take part:

LOS ANGELES: You, with the glasses and the raincoat, what's your name?

LOS ANGELES: You!

NEW YORK: Me?

LOS ANGELES: You.

NEW YORK: Matthew.

Hole in Space thus at once captivated (and captured) participants and fostered a range of self-consciousness relations—social, spatial, and technological.

To complicate their experiment and enhance the potential for such self-consciousness, Mobile Image designed *Hole in Space* as a succession of different experiences unfolding over its three-day course: a day of spontaneous, unexpected encounter; a day of more deliberate participation; and a day of mass use. Passersby who chanced upon the work on the first day had to figure out what it was and how it functioned, and, as already mentioned, the behaviors that emerged on that day mainly consisted of rudimentary exchanges, tests, and explorations. As the artists explained:

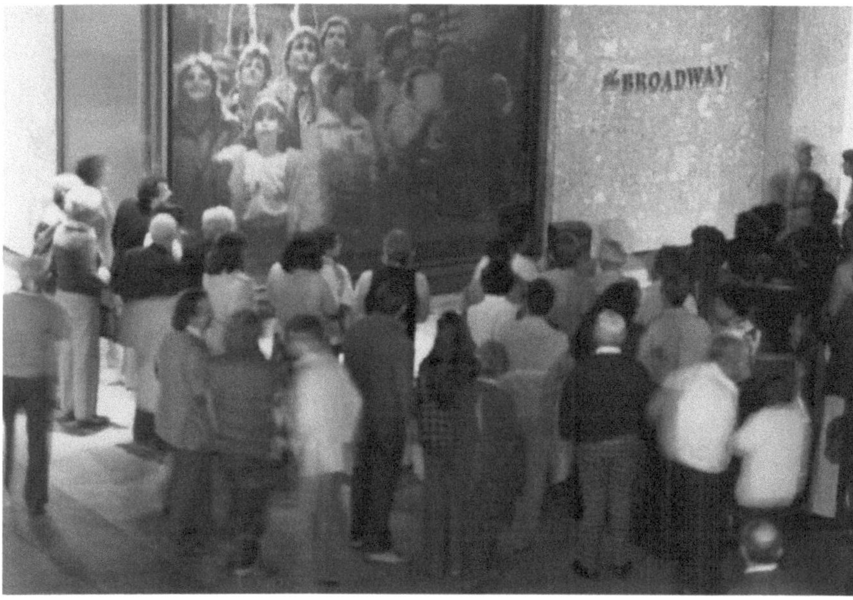

Curious pedestrians must discover for themselves, first that the picture on the screen is not a film or videotape, but a live television image. Second, that it is coming from across the country. Third, that the people they see on the screen are also looking at a screen in a window. Fourth, that the people they are seeing and hearing are seeing and hearing them—that in fact what they have stumbled upon is an open channel, a live two-way television link through which they can see, hear and talk with people three thousand miles away, almost as if they were standing on the same street corner.[17]

A day off was scheduled in between the work's first and second sessions to allow word to spread and people to plan their future attendance and coordinate meetings with friends and family. This resulted

2.2

Mobile Image, *Hole in Space (A Public Communication Sculpture)*, 1980. Courtesy Rabinowitz/Galloway Archive.

in larger crowds on day 2, a combination of newcomers and returning participants—the latter ready for the experience and able to explain to the former what was happening and how it worked. The social dynamic also became much more organized, with "highly structured, highly coherent games and conversation," as Lifton-Zoline observed, including charades and group songs.[18] The work began to generate its own "popular" customs and history. "It developed its own sort of humor and gags that people would play on each other," Galloway explains. "It developed its own myths, stories, things that were said to have happened that didn't."[19] Mobile Image had also announced the project to local television stations in advance, encouraging them to cover it but strategically telling them that it would begin on what was actually the work's second day. Media coverage could thus serve as widespread publicity for the final day without spoiling the initial period of discovery on day 1 and the word-of-mouth, still relatively intimate, quality of day 2. As Rabinowitz explained, "We didn't use the media just to toot our horn. We used them to help create the project by revealing the different levels of exposure, the different levels of acculturation and acclimation within the culture."[20] The media thus became part of the work, both an instrument of publicity and a featured element—and another acknowledged part of the broader apparatus. While covering *Hole in Space*, the TV crew from New York's ABC News were themselves incorporated into the cross-country transmission: the reporter, Peter Bannon, stood at the edge of the crowd, practicing his lines ("Interactive sculpture? Or is it satellite sculpture?") and recording multiple takes, while people in Los Angeles could see him and his news camera pointing at them from the opposite coast (figure 2.3). At one point, Bannon shouted to the LA crowd "You're on TV," to which they responded, "No, *you're* on TV."[21]

The much bigger crowds that congregated on the final day were largely a consequence of that news story, which aired later that night: "Tell your friends," Bannon told his television audience, "it'll be up again tomorrow night, Century City in Los Angeles, Lincoln Center here in New York . . . New York time will be 8:00–10:00pm." Mobile Image described this final session as "over-democratized bedlam,"[22] in which the previous sense of collectivity broke down, group activities and identities generally supplanted by multiple parallel encounters between individual friends and families, planned in advance. People crammed into the space, shouted

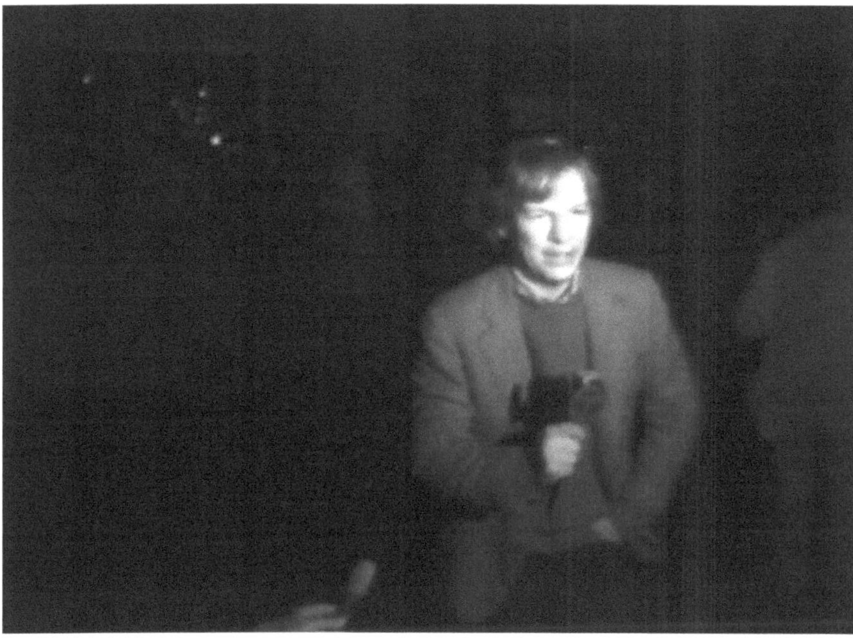

out for their acquaintances, excitedly said hello and briefly caught up, some seeing each other for the first time in years; new partners and babies were introduced; some even brought signs to beckon a particular person. All of this happened concurrently, the cacophony heightened by an overload of audio feedback. Whereas this chaotic scene, what Lifton-Zoline characterized as "a loud . . . non-cooperative texture of familial display,"[23] can be understood as a retreat into the kinds of personal interactions from which the previous days had markedly diverged, *Hole in*

2.3

Peter Bannon, ABC News, reporting from *Hole in Space (A Public Communication Sculpture)*, screenshot from video documentation, 1980. Courtesy Rabinowitz/Galloway Archive.

Space continued to blur the boundaries between private and public, here essentially bringing private communication out into the open for public consideration. As Rabinowitz observed, "People walked away saying 'You know, listening to them talk to their mothers, that's how I talk to my mother!' To see your own intimacies as part of a whole was very powerful. It was a different kind of unifying experience than the other two nights. It had a tribal feeling."[24]

In these ways, Mobile Image engaged the urban public as a space of contention, complexity, and ambiguity, whose political and technological progress depends on the organization of its parts. This meant providing people with an experience that was not only astonishing, immersive, suggestive of new potentialities, but which also shed light on the myths, limits, and commodification of technology as public tool. *Hole in Space* presented (and represented) the public sphere as a battleground, subject to power and perception, and most viable when conceived as multifarious and continuously shifting. It was crucial at once to evoke pervasive ideals of technologically enabled sociality and to underscore the specificities of time and place—the particular (historical) moment and location, the context—that may complicate, if not contradict, such ideals. Participants could then begin to see the personal and the social, the political and the technological, as intertwined, the public and the private not as mutually exclusive but rather as interrelated sites of experience—and struggle. This relationality could then form the basis for potential transformation.

Hole in Space thus raised a range of vital questions that became ever more pressing over the ensuing decades: How has the city (and public space in general) changed, and what is the relationship between such change and the emergent technologies of the time? What constitutes "the urban" in a world of mass media and worldwide networking? What myths persist, and what realities—new and old—have to be grappled with? What are the possibilities for a technologically enabled arena for progressive, accessible, and joyful communication between diverse populations, a heterogeneous arena that is neither fully subject to commercial control nor overly reliant on idealized, nostalgic notions of the public sphere? How might activist efforts—works of art included—productively reflect the effects of segregation, division, and privatization, engaging rather than attempting to conquer the fragmentation of the self and the community? And how might such efforts make use of new telecommunication devices while also remaining conscious of the commodification of urban experience and communication, and without simply strengthening the cult of new gadget? *Hole in Space* belongs to a specific historical and art-historical context in which urban public space—its inclusions and exclusions, its underlying economic, technological, and ideological structures—emerged as a primary site and subject matter. The work's arrangement of mechanical devices, urban spaces, and bodies must be understood in this context, while its evocation and explicit use of screens set within shopwindow displays belongs to emergent discourses on the relationships between mass media and social production, the struggle over public encounter and engagement, perception and self-perception, and attendant conceptions of agency, participation, and activism.

PUBLIC/PLACE/SITE

Constructing a public telecommunications sculpture in New York and Los Angeles in 1980 meant engaging the urban as a complex socioeconomic and aesthetic struggle over place and site. It would be easy to posit *Hole in Space* as part of the then-popular lament of the decline of the city as the arena of idealized public encounter. The work would thus be understood as an artistic intervention, at once projecting and legitimizing a longing for the polis assembled in squares and cafés, unencumbered and egalitarian, for a citizenry united in the labor of social progress. A number of writers and critics at the time—from nostalgic liberals to Marxist modernists to postmodern urbanists who shaped much of the discourse around the aesthetic politics of urban social formations—echoed Jürgen Habermas's assessment of the structural transformation of the public sphere through capital and new technologies, understood as almost unilateral forces of ever-expanding exploitation, privatization, and fragmentation. The "explosion" of mass media, what Frederick Williams would, in 1982, term the "communication revolution," was observed wearily as a main contributor to the pervasive psychological-geographic disorientation, if not all-out crisis, that especially afflicted the centers of Western society and the "urban" as public.[25] Around this same time, Paul Virilio articulated a more complex if equally dystopian connection between city, public, and technology as both physical and ideological (or imaginary/imaginative) elements within the context of "an imminent 'postindustrial' deurbanization," a "deregulation of the management of space [coming] from an economic and political illusion about the persistence of sites constructed in the era of automotive management of time." The "polis" is thus replaced by the screen, which broadcasts "the last image of an urbanism without urbanity . . . where tact and contact give way to televisual impact."[26] Although certainly insightful, such mournful critiques are generally based on a concept of the public sphere as a unidimensional, total organization of experience predicated on what Miriam Hansen calls "formal conditions of communication (free association, equal participation, deliberation, polite argument)."[27] This leaves much urban "public art" to function as a symbolic assuagement of the lack of such conditions and the resulting loss of wholesome collectivity.

In contrast, the contemporaneous writings of Manuel Castells provide a more adequately nuanced context for a work such as *Hole in Space*. Castells's sociological-empirical studies contain some of the most productive assessments of the relationships between "high technology," the "modification of the urban form," and new spaces of production.[28] These studies provide a basis for understanding how technological and urban development in the United States during the 1970s and early 1980s, and its effects of division and privatization, can be critically reflected and engaged by artistic activism, in ways that do not rely on regressive, and ultimately ameliorating, end-of-public-space narratives. Was (and is) there a public potential for advanced telecommunications to provide a progressive, counterpublic experience, one that is not totalizing—either in its dream of unity or its negative critique of uniform fragmentation—but productively heterogeneous, complex, and ambiguous?

According to Castells, technology and technological production alter urban, national, and global relations through capital and information flow, producing privileged hubs or "nodes" within cities, regions, and beyond. Technology—among other factors such as demographic trends, cultural patterns, new neighborhood movement, and the emergence of women and minorities as political forces—transforms urban space: on the one hand, it rejuvenates and animates; on the other, it segregates, decentralizes, gentrifies. Telecommunications and automation allow for the outsourcing of manual-industrial production, controlled and managed remotely, leading to what Castells calls "interregional division of labor," while the transformation from the production of physical goods to an information and service economy stimulates a revitalization of "central business districts" and a significant, if limited, gentrification of adjacent neighborhoods.[29] The "electronic home" and the "electronic office"—complete with actual and potential online information services, including "electronic mail, electronic banking, or teleshopping"—fuel territorial sprawl, both urbanization and suburbanization, as well as an individualization of socio-spatial relations. The "wired" or telecommunicated city turns places into "unifunctional units," reduces public space into spaces of leisure, "for those who have the time and money," and spaces of wandering, "for those who do not fit into the functional assignment of work and residence."[30] Homes become at once "instant receivers of planetary information and personal refuges of selective consumption," as not only does the personalization of communication experience enable people to stay home, but also "they will be able to leave home without leaving their inner experience." Although filled with images and sounds and news from all over the world, private experience and its space are increasingly delocalized, capable of being switched on and off, selected on demand, dissociated from neighbors and neighborhoods, from the city as local network of mediating sites.[31] "Choice," Castells explains, may mean that, from a technological point of view, there is "no more mediation between the individual and a global culture . . . thus in between there would be no more society and no more city."[32] However, that does not mean that there aren't other experiences, other sites; the city and its "public" are not "dead" but rather are actively transformed and contested, as meaning production within it is multiplied, assigned along different connections and hierarchies, some oppressive, some potentially emancipatory.

Whereas flows of information and capital may directly and seamlessly connect the most privileged nodes of the private, including business as well as leisure and dwelling, the system imposes a value system marked by such connectivity, superimposed on other suppressed and excluded layers of psycho-geographic and socio-communicational urbanism. Studies of the evolution of New York and Los Angeles conducted in the early

1980s have shown several simultaneous processes of polarized growth that created "distinct social spheres," which nonetheless are and need to be linked within the same functional unit: the city. A kind of "urban schizophrenia" results, as myriad new social movements, neighborhood alliances, grassroots political organizations, service workers, transit hubs, squats, shelters, and other formations occupy and move through the sites and sights that made up the urban as interdependent, evolving constellation.[33] This development of the new "supercity" is violently uneven in many regards, but Castells argues that it nonetheless bears the potential for resistance, especially if and when the "simultaneous life and death of our great cities" refers as much to the functional-material conditions of city life as it does to the very concept, ideal, and ideology of the dynamics and experience of the urban. With this attempt to change the perspective on what constitutes a viable urban public comes the need to differentiate: discussions about the decline of urban space and urban flight, revitalization and new publics, must acknowledge the ways in which the uneven development of space through capital and technology affects particular industries, communities, classes, genders, and ethnicities differently. Depending on how the experience of this development is organized and communicated, it can further entrench or realign positions of agency relative to privileged nodes and flows. Without justifying or putting a false-positive spin on urban blight, as some populations move out, others move in, establish communities, attitudes, and territories, struggle, and live.

Responses to this managerial-technocratic spatial reorganization and delocalization of experience, to the tendency toward the de-connection or increasing abstraction of the relation between people and the sites wherein knowledge and imagination are produced, likewise take a variety of forms, with conflicting urban meanings and functions, from a variety of urban actors.[34] These reactive and proactive disruptions develop along the very logic imposed: the new economy's delocalization and decentralization of territories and tools is met with desires for localization, cooperation, and community; the "legitimation crisis" of the democratic nation-state in light of its protection of global capitalist interests and subsequent privatization or outright withdrawal of welfare and social services is met with demands for and a rise in local autonomy and self-management.[35] According to Castells, whether these tendencies will ultimately lead to regressive reproductions of hierarchies and mechanisms of competition and exploitation or to social and economic innovation depends on the creative (re)construction of communication in urban environments. What is needed, however, is more than just increased access to new media gadgets. Despite the rising decentralization of broadcasting and a decrease in the homogeneity and monopoly of the messages sent out over the ether, "we are facing a communications disaster," Castells warns, as most so-called technological progress in telecommunications is reduced to either one of two related models: in the first, the information apparatus is diversified according to the needs of an increasingly segmented market, leaving intact the "monopoly of code" and the one-way logic of the broadcast; in the second, the availability and adoption of technological know-how comes without the productive transformation of "social patterns of communication," leading to a cacophony of individuated voices and microcultures that lacks the strategic reorganization of collective political agency based on alliances of necessity and shared reference (without adhering to traditional, bourgeois public sphere notions of consensus, unity, and harmony).[36] The most urgent need is therefore the ability to forge new connections, to designate to communications technology a new collective use value that can organize the experience of a heterogeneous public.

Castells's crucial contribution to the discourse about urban space as public space thus consists of not only an urgent plea for the revitalization of streets and squares as sites of critical-productive exchange but also a historicization and politicization of the public sphere as a concept, one that determines hegemonic as well as potential counternarratives of how social formations, experiences, and agency are formed and exercised—and, in turn, the function and usefulness of artistic, technological, and civic activism. During and since the height of the debates about urban development and public art in the postindustrial city in the 1980s, authors such as Rosalyn Deutsche, Craig Owens, Martha Rosler, and Miwon Kwon have questioned the "naturalization" of the history and ideology of the public sphere and art's role within it.[37] At stake in this politicization is, as with Castells, a history of urban public space as multiple, contested, and unstable, and hence as potentially transformative. The democratic potential of social encounters is measured not by their ability to revive some mythic sense of authentic urbanity but rather by the capacity for voices, experiences, and constituencies to change and multiply, seek new alliances, and form new publics, expressions, and visibilities. As Deutsche argues, to avoid the politics of the public sphere's historical construction beyond, and in contentious relation to, its bourgeois ideals is to forgo the possibility of social progress and to fall back on violent nostalgia. "Within this idealizing perspective [of the traditional city]," she explains, "departures from established spatial arrangements inevitably signal the 'end of public space.' Edge cities, shopping malls, mass media, electronic space become tantamount to democracy's demise."[38] This conservative appropriation of public space denies the potential of new and divergent modes of traffic and travel, of orientation, access, and encounter. It relegates so-called peripheral experiences and demands—be they transitory and transitioning, "inner-city" housing and health needs, or the desire for representation and presence—to the "private," outside the purview of "public interest," while legitimizing the expansion of private (economic, business) interests to control, police, and "revitalize" areas of "urban decay" in the name of public safety and civic stability.

Deutsche detects an aversion to a politics of heterogeneity and ambiguity in Zeitgeist analyses that aim to connect questions of space, culture, and aesthetics in postmodern society. Although she finds much to laude and admire in the discourses that shaped major strains of thought and artistic production during the 1980s—epitomized by the works such as Fredric Jameson's *Postmodernism, Or, the Cultural Logic of Late Capitalism* (1984), David Harvey's *The Condition of Postmodernity* (1989), Edward Soja's *Postmodern Geographies* (1989), and Michael Sorkin's anthology *Variations on a Theme Park: The New American City and the End of Public Space* (1992)—Deutsche is troubled by the extension of modernist tropes and norms that leave little room for an innovative transformation of the sites and means that organize experience, much less the very concept of experience itself.[39] If late capitalism's increased centralization of power and production manifests in an experience of alienation, displacement, and disorientation, is the yearning for a position of unity, of a perspective from which to map coherently an ideological totality, merely a reproduction of the mechanisms that create the desire for such unity in the first place? In other words, is it not only possible but also necessary to reconfigure and reconstruct the space of politics—a new politics of perspective, encounter, interaction, and thus individual and social subjectivity? The "totalizing impulse" that Deutsche finds at the root of the desire to mark a privileged and unconflicted place of the public subject, or public *as* subject, manifests in an indifference to difference, feminist and otherwise.[40] The panoramic vantage point is itself product and site of the contradictions and struggles of (post)modern positionality. What, then, is the role of art as public in urban space beyond a symptomatic affirmation of the dichotomy of valorizing fragmentation and exclusive totality, of symbolic-popular ownership of perceptual and social movement, on the one hand, and a prescriptive-analytic unity of ideological certainty, on the other? Can the very concept of the public and its sphere of production be reconfigured rather than disbanded or resuscitated, and hence be understood and utilized as a tool as well as a product, as a space and technology that determines the identity and outlook of social formations and is itself constructed by such entities and thus accordingly affirmative or potentially transformative of them?

PUBLIC/ART/CITY

Mobile Image engaged these very issues by establishing a technologically enabled mode of public gathering that was at once emphatically local and explicitly designed to transcend the local, that underscored both connection and difference, and that simultaneously satisfied needs and threw them into question. Although the work was set in two very different cities, with different meanings and functions of public space, different physical and cultural landscapes, and different local art scenes, it fit within the context of each, as artists and arts organizations in both Los Angeles and New York were struggling with the politics of urban publicity at this time. *Hole in Space* was temporary, unannounced, and unidentified as "public art." Yet, by assembling people on the street and enabling them to produce social arrangements both in that place and through the screen, while explicitly incorporating markers of broad cultural operations—the performing arts complex, the shopping district, the television news media—the work highlighted the material and immaterial relations on which notions of public space, experience, and art depended. In addition to repurposing devices typically associated with entertainment spectacle or corporate telecommunications, it reconfigured particular sites and technologies that were relatively new but already standardized according to preestablished functions and symbolic meanings based on nostalgic public sphere ideals.

Both Lincoln Center and Century City were the products of urban development and gentrification efforts of the 1950s and 1960s that sought to create inviting places for public activity, whether it be attending a performance or strolling through an open-air shopping mall. Such efforts were renewed in the 1970s and 1980s, as cities attempted to jumpstart their sputtering economies through major revitalization projects, again via private–public partnerships. As urban historian M. Christine Boyer has described, following the mid-1970s fiscal crisis, cities increasingly replaced holistic urban planning with uneven development, fueled by zoning accommodations and tax incentives that facilitated huge real estate–driven construction projects. "In return for these development gifts," Boyer explains, "the private developer must provide some of the amenities generally lacking in the public realm." The result is an "illusionistic" city, in which luxury districts mask the neglect of interstitial places and society "call[s] something public when it is indeed not."[41]

Along with an array of newly constructed or renovated spaces—cultural complexes, pedestrian-friendly marketplaces, charming neighborhoods—founded upon romantic images of city life, public art became a key component of this so-called revitalization process. Large-scale sculptures by canonical modernists such as Pablo Picasso, Alexander Calder, and Henry Moore were commissioned for corporate or governmental plazas, often through "Percent for Art" programs that popped up in cities across the country. Largely funded by the General Services Administration's Art in Architecture Program (established in 1966 and fully activated in 1972), such works—subsequently derided as "bureaucratized art" or "plop art"—were meant to counteract the perceived deterioration of urban public spaces, seen as havens of homelessness and crime, and more generally as a way to mobilize art as a tool of urban development and redevelopment. These sculptures would enhance the "symbolic economy" of the places in which they are situated while ostensibly establishing what Tom Finkelpearl calls "the city as museum"—that is, a place for civil discourse via reasoned contemplation of autonomous objects set in open areas.[42] At a moment when public funds were increasingly being used to prop up urban spaces that were already partially if not fully privatized, public art was being used to help re-present those spaces as loci of a reinvigorated public sphere.

A range of more eccentric and participatory public art projects also emerged in the 1970s and early 1980s, seeking to either provide unconventional experiences in otherwise conventional spaces or occupy more peripheral locations. Organizations such as the New York–based Creative Time funded various works in which corporate and municipal spaces—office-tower lobbies, parking lots, historic government buildings—were temporarily transformed into sites for public gathering and multimedia, participatory experience.[43] Some attempted to establish alternative spaces not just for the general art public but also for specific geographic, socioeconomic, ethnic, gender, or racial constituencies historically excluded from—if not forcibly displaced by—urban development and revitalization enterprises. Neighborhood-based art projects, community arts organizations, and artist-run venues proliferated nationally in the 1970s, intended to provide such constituencies with a reasserted sense of place, with needed physical and discursive forums for public expression, communication, visibility, and sociality. As part of the more general "alternative space movement," numerous exhibition spaces founded on shared identities emerged. For some artists, the production of alternative public space and social experience in the urban realm was not a by-product of their work; it was their art. In the early 1980s, for example, the art-activist collective Group Material opened a storefront gallery in a low-income, largely Latino neighborhood in Manhattan's Lower East Side, featuring socially informed exhibitions that were meant to create a space—or, as Richard Goldstein put it in an early review, an "anti space"—for ethnically and culturally diverse local publics to come together, communicate, evaluate, and bond.[44] These ventures, along with much of the alternative art space movement, can be understood in terms of what Lucy Lippard calls the "lure of the local," in which the embrace of a particular place by resident "insiders" provides a forum for marginalized identities and serves as an therapeutic assuagement of contemporary placelessness, a response to the intensified mobilization of bodies, information, images, and commodities, on the one hand, and the increased homogenization of urban space, on the other. The local becomes a hopeful antidote to the fragmentary, alienating experience of a society at once decentered and standardized by forces of capitalism.[45]

Yet, these attempts to democratize the public sphere via emphatically localized sites of assembly and discourse can be understood as early iterations of a trend that Miwon Kwon would later criticize as fundamentally nostalgic and binaristic. Such endeavors, Kwon argues, seek to retrieve and resuscitate romantic notions of urban public life—vernacular sociality, face-to-face exchange, a sense of belonging, of self, a way to feel secure, empowered, and "whole" again—and are dependent on a rigid opposition between place and placelessness. "Conditions of groundedness and connectedness are themselves imagined as resistant to the forces of the dominant culture."[46] However, the desire for a place of one's own—for the experience of local specificity, cultural authenticity, identity, and difference—is no less a central component of capitalist expansion than the seductive allure of mobilization provided by modern technologies. Whereas "alternative space" movements approached urban space as a site of struggle and attempted to render it more heterogeneous, they arguably ran the risk of succumbing to the problems Kwon identifies. Recognizing that risk, some groups (Group Material among them) began to rethink their practices in the early 1980s, rejecting the gallery space and shifting to "guerilla" tactics targeting non-art locations (streets, city walls, the transit system) and audiences (commuters, passersby).[47] The challenges of space and place, however, extend beyond the question of art institutions per se to the fundamental problems of the local and the authentic. Relying on Henri Lefebvre's insights into the dialectical rather than oppositional relationships between "the processes of expanding abstraction of space and the production of particularities of place," Kwon advocates for practices that consider these seeming contradictions and our contradictory desires for them together, as mutually sustaining relations: "How do we account for, for instance, the sense of soaring exhilaration *and* the anxious dread engendered by, on the one hand, the new fluidities and continuities of space and time, and on the other hand, the ruptures and disconnections of space and time? And what could this doubleness of experience mean? In our lives? In our work? Within ourselves?"[48]

Predating Kwon's words by two decades, *Hole in Space* elicited precisely these questions, as did other works from the period that similarly approached urban public space reflexively, dialectically. Such works not only provided novel social experiences but also fostered encounters with the functional-material qualities of such space and experiences. They offered at least the potential for audiences to consider the relationship of those qualities to larger power structures, technological systems, and media apparatuses, provoking essential questions about public space in the contemporary era: How it is defined and constituted? What is its purpose and function? How it is experienced?

These questions emerged in relation to then-current economic decline and uneven recovery, along with profound shifts in the image of the contemporary city—be it termed "postindustrial," "late-capitalist," or "postmodern"—via both its representation in mass media and changes to its actual physical structure. As urban sociologist Sharon Zukin points out, whereas photography and film had been determining how urban space is imagined, understood, and "sold" for decades, beginning in the 1970s, with redevelopment increasingly turning to consumption activities and urban space increasingly being reconfigured to accommodate and promote such activities, "the material landscape itself—the buildings, parks, and streets—[became] the city's most important visual representation." Cities have always been founded on a symbolic economy that "defines what—and who—should be visible and what should not, [based] on concepts of order and disorder, and on uses of aesthetic power," Zukin explains. "What is new about the symbolic economy since the 1970s is its symbiosis of image and product, the scope and scale of selling images on a national and even global level, and the role of the symbolic economy in speaking for, or representing, the city." The result was the arrival of aestheticized, gentrified, and privatized spaces—what Zukin calls "sites of visual delectation"—populated by restaurants, stores, and other entertainment, or retail-based places for middle-class people to stroll and consume, along with much broader changes to the overall material and demographic makeup of the city: "Developing the city's symbolic economy involves recycling workers, sorting people in housing markets, luring investment, and negotiating political claims for public goods and ethnic promotion."[49] Lincoln Center was an early harbinger of such transformation. Conceived in the 1950s as part of Robert Moses's revitalization program and constructed in the 1960s, the complex co-opted a large residential neighborhood marked as "urban blight," displacing its residents and converting the area into an anchor for upscale real estate development generated by the symbolic value of a massive performing arts center. For the first time, "the allure of culture legitimated the use of urban renewal funds by the city's growth machine," Zukin explains. "While building Lincoln Center gained New York City international prestige, on a local level it confirmed the cultural hegemony of the city's postwar elite."[50] With such projects, New York began to transition from America's quintessentially "panoramic" city—a supposedly coherent, rationally planned space founded on modernist ideals of form and function—to what M. Christine Boyer terms the "City of Spectacle," which bloomed around 1980, when "the transformation of the material world by invisible bands of electronic communication encircling the globe, by computer-simulated visual environments, and by theatricalized image spectacles seemed by extension to have decomposed the bits and pieces of the city into an ephemeral form."

Los Angeles, Boyer contends, epitomizes this new paradigm—the city as a series of pictorial tableaux, of imposed images divorced from history and memory, a scenographic "nonplace" where nostalgic visions of urban public space and experience serve to reinforce dominant power structures.[51] As Reyner Banham observed in his early 1970s survey of LA architecture, a recently developed neighborhood such as Century City was essentially a "planned illusion," centered around the shopping mall, which offers a commercial setting to satisfy the desire for pedestrian spaces in a culture defined by the automobile. Representing "the newest fantas[y] of the good life in Southern California," Banham explains, such spaces are Los Angeles's "alternative to Main Street, the natural foci of a highly mobile population that measures distance in time at the wheel."[52]

Venturing out into the urban environment at this pivotal moment, certain artists enacted direct, politically pointed interventions in quotidian public places. Rather than occupying architectural space or vacant lots for prearranged, self-contained art events or establishing alternative sites for exhibition, these artists produced performative works that were often unannounced and set on the street amid random passersby, meant to reclaim and reframe public space momentarily in ways that could politicize its otherwise overlooked routine function. In Los Angeles, beginning in 1976, Kim Jones performed *Mudman*, in which the artist covered his body in sticks, mud, and other substances and affixed himself to the city's architecture or wandered the streets as an alien "other." In 1978, Senga Nengudi produced *Freeway Frets* (figure 2.4), in which she and several collaborators transformed marginal urban areas—freeway overpasses, traffic islands—into settings for art, performing rituals that drew attention to the relation between urban space, social structures, and racial, gender, and geographic identities.[53] Earlier in the decade, the Chicano art collective Asco began producing a kind of absurdist public theater that similarly linked the politics of space and identity, often involving the use of their own bodies, juxtaposed against the specter of urban development (figure 2.5). As will be discussed in chapter 3, their East LA community was a potent backdrop—a recent victim of urban revitalization programs, in which the construction of new highways and retaining walls divided formerly connected neighborhoods. Part of what has been called the "expressway generation," Asco's members were keenly aware of how public policies and city planning creates conditions of disparity.[54] While continuing the public-art focus of the "Chicano arts movement," the group broke with the types of activities most closely associated with that movement, particularly muralism, which they saw as conservative in both iconography and function. Asco's goals were not only increased visibility and inclusion—frequently the aim of "authentic" activist activities, community murals among them—but, as Amelia Jones explains, "discussion, contestation and open-ended self-performance *even in relation to the artists' chosen coalition identity*."[55] Their public interventions can be understood in contrast to both dominant (white, middle-class) culture, with its imposed images of normative urban publicity, and their own community's traditions of resistance. Throughout the 1970s and early 1980s, Adrian Piper similarly treated the public sphere as a zone of social confrontation, the experience of which depends on one's performed and perceived identity and status, but which also holds the potential to redefine relations. Piper mainly sought that potential on the personal level via works that engaged individuals, challenging their prejudices and linking those prejudices to entrenched histories.

Artists such as Jones, Nengudi, Piper, and Asco approached urban space not as a utopian sphere of individual and collective expression but

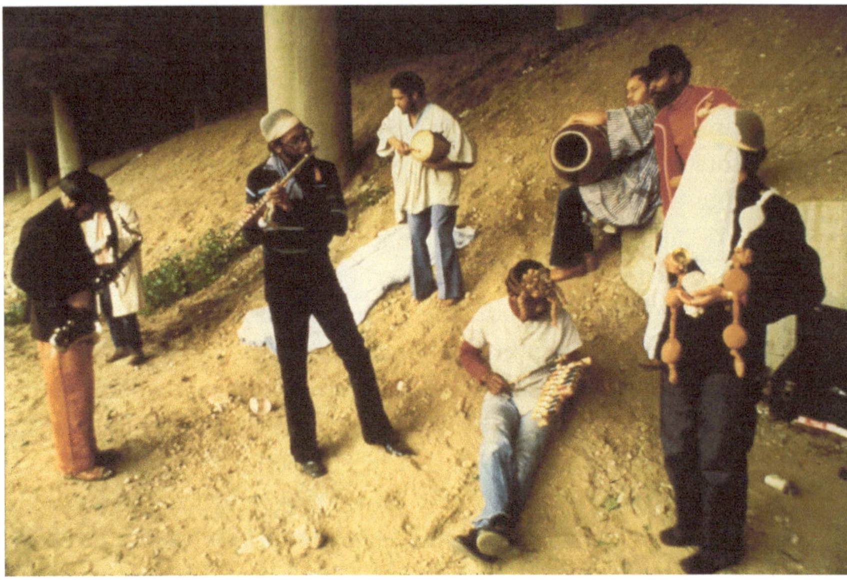

rather as an unstable site of conflict—both intra- and extra-communal—while also suggesting new social, economic, racial, and geographic arrangements. Although diverse in artistic and political intention, they, like Mobile Image, understood such space not as some "pure" public sphere but rather as always a site of struggle between the ideal and the material, one that produces ever-evolving sites that themselves facilitate the performance of identity—social and individual, self and other—in and through space. Unlike *Hole in Space*, however, the aforementioned works were less open-ended and reliant on public participants and, more pointedly, issue oriented. Even when observers were unaware of the particular artist's or collective's identity, or when unsuspecting audiences were encouraged to join in, these artists focused their work on particular topics—race, class, otherness in general—enabling it to be more explicitly political

2.4

Senga Nengudi, *Freeway Frets*, 1978.

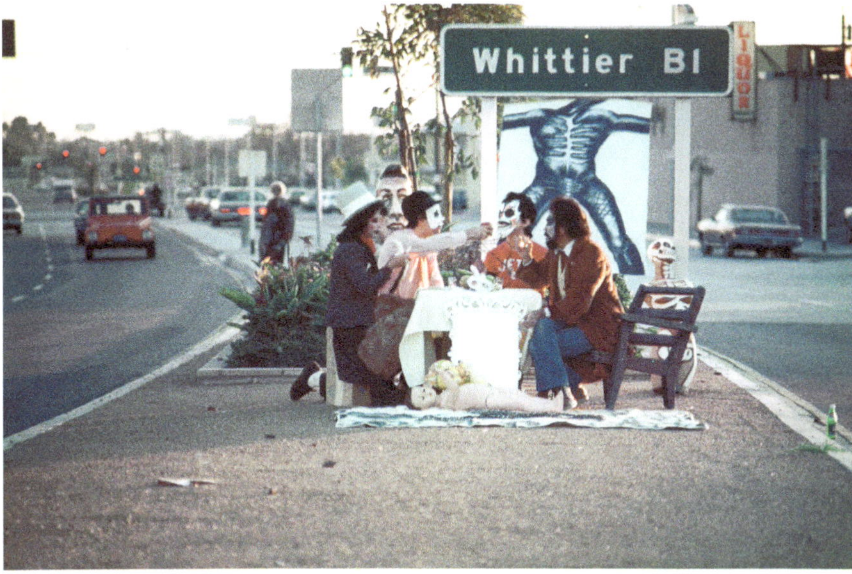

while also relying on arbitrary encounters and allowing for myriad interpretations. These artists already seemed to understand what would later become a central tenet of Mobile Image's work, as manifested in *Electronic Café*: that the public sphere constitutes and is constituted by very specific inclusions and exclusions, that it is the site of very particular struggles.

Other artists produced works intended to foster reflexive, open-ended reconsideration of spaces, bodies, and histories ordinarily marginalized—or rendered invisible—by dominant public sphere models and to expand conceptions of the urban "public" to include broad social, economic, and technological systems that enable those exclusions. In 1980, for example, the New York–based collective Colab mounted "Real Estate Show," an open-call exhibition in a vacant building in the Lower East Side. At a moment when real estate prices were beginning to rise and the area was

2.5

Harry Gamboa, Jr., photograph of *First Supper (After A Major Riot)*, performance by Asco, 1974. © Harry Gamboa Jr.

marked for redevelopment, resulting in a rash of evictions throughout the neighborhood, the show was, in the words of participant Ann Messner, "a provocative stance deployed to expose the city's nefarious relationship to . . . the urgent concerns of an impoverished community."[56] The contemporaneous New York–based work of Mierle Laderman Ukeles likewise focused on the larger public apparatus—in her case the seemingly apolitical, mundane municipal spaces, systems, and labor that sustain city life. In Los Angeles, artist Suzanne Lacy similarly attempted to expand conceptions of public space and institutions to include broad systems and diverse and at times conflicting perspectives. Works such as the 1977 *In Mourning and in Rage* (organized in collaboration with Leslie Labowitz) transformed a public space into a memorial to victims of violence against women while implicating the broader media apparatus, while also demonstrating how both could function as sites of public action, potentially reorganized into platforms for intervention, dissent, and the assertion of otherwise silenced voices.[57] Here, as in the aforementioned work of Ukeles, Colab, and others, "public art" becomes at once a pointed issue- and site-specific activity and a broader meditation on—and politicization of—the very definition of public, the various systems that comprise it, the contest over its conceptualization, function, inclusions, and exclusions, and the heterogeneous relations that sustain and are sustained by it.

Hole in Space similarly incorporated an expanded definition of urban public, conceived as a relational site of struggle, while also modeling a technology of assembly and communication that could both elicit critical reflection and potentially restructure psycho-spatial arrangements and their underlying social, cultural, and aesthetic politics. And, like Lacy and Labowitz, Mobile Image approached the increasingly ubiquitous technological and media apparatuses as public sites in and of themselves—sites that were in the process of being privatized but which nonetheless remained very much entangled with public existence. *Hole in Space* was, however, distinctly improvisational and participatory, not only involving others and dissolving the boundaries of the work of art but also capturing arbitrary passersby who could then make the experience their own, inviting them to discover, produce, and innovate, to not only connect with random strangers across the country in real time but also arrange future uses over the subsequent days. The work expanded definitions of urban public to include the technological apparatus through which people communicate, move, extend themselves; technology was posited as fundamentally "public," while the experience of the physical city was posited as fundamentally technological. *Hole in Space* belonged to a moment in the history of public art when the transitory was posited as a corrective to the seemingly staid conventionality—and reifying effects—of large-scale, permanent sculptures set down in officially designated public spaces. As Patricia C. Phillips argues, such temporary, site-specific works enabled the artist to engage the public without "mandating the stasis required to express eternal values" that suggest a monolithic audience despite a recognizably fragmented society.[58] Yet, as Kwon argues, impermanence and uncertainty alone are not enough; the indiscriminate embrace of such qualities potentially validates, even romanticizes, the material and socioeconomic realities of an ever more mobile, transitory, and dislocated society.[59] The temporary and site specific cannot simply serve as antidotes to the static and generic. Again, effective public art must treat public space as contested, and it must engage that space dialectically.

Hole in Space placed various spatial and technological arrangements into dialogue. At a time of industrial decline, financial speculation, and urban revitalization, the work approached the public sphere as a site—physical and electronic, material and immaterial—where history and politics determine both hegemonic and potential counternarratives, formations, and experiences. Consequently, the work challenged a range of entrenched binaries—here and there, real and virtual, variation and homogenization, individual and collective—while also suggesting a dialectical view of several crucial struggles of the day: tensions between globalization and decentralization, on the one hand, and the desire for localization, cooperation, and community, on the other; the function of technologies of mass communication that provide highly individualized experiences but which, as Sharon Zukin argues, also impact how cities are imagined, commercialized, and constructed; the uneven development of space, what Manuel Castells refers to as the "simultaneous life and death of our great cities," and the different effects of that disparity on particular communities and peoples; and ultimately society's ever more obsolete distinctions between so-called public and private spheres. The specific places linked by *Hole in Space* were crucial. Archetypes of mid-twentieth-century urban development, Lincoln Center and Century City epitomize the privileged hubs or "nodes" that, according to Castells, result from flows of capital and information shaped by technological development. In Los Angeles, the crowd was situated inside such a node, amid the outdoor mall's aestheticized urban space, in front of its "anchor" department store. The New York location was, by contrast, outside the node, just beyond the cultural complex and adjoining square. (The work's participants were consequently quite different, at least initially: those in LA were mainly patrons of the shopping district; those in New York were not theatergoers but rather pedestrians who just happened to be walking by.) *Hole in Space* thus established a relationship between such nodes and the spaces between them, effectively politicizing both: the new urban plaza, so frequently lamented as having been depoliticized, as a commercial bastardization of a once-effective public sphere, and the urban street, often either idealized as a vital site of democratized sociality, of communal interaction and expression, or simply seen as apolitical, outside the hub, merely a passageway through which people move. As Rosalyn Deutsche argues, identifying new and divergent modes and uses of public space requires the rejection of nostalgic public sphere ideals and lamentations of their demise, along with a refusal simply to dismiss so-called peripheral places—in between, commercial, or electronic, material or immaterial—as outside the public interest. Mobile Image at once eschewed such ideals and recuperated those places as potential sites for collective activity and for the perception and navigation of the relationships between history, self, and the social, of "public" itself, its possibilities and limits.

In its image of contemporary sociality as taking place in and through the screen, as always already mediated, *Hole in Space* can be understood in terms of Boyer's notion of the City of Spectacle, a physical space transformed by spectacular representations and telecommunications systems into an ephemeral form, a nonplace. Yet, unlike many contemporaneous diagnoses of the postmodern city, the work did not cast these spaces as simply "faux" urbanism, as simulacra, as drained by mass media technologies of all meaning and shared memory. The critical-productive practice of Mobile Image was not limited to exposing false images without hope for potential transformation. *Hole in Space* was an experiment in whether there is at least a potential for advanced telecommunications to enable a heterogeneous, complex, counterpublic experience with the potential to critique, unify, and transform. Without resorting to totalizing views, whether utopian or dystopian, the work both reflected the current situation and modeled a way to use such technologies productively, validating new sites and modes of assembly and communication without claiming them as any more inherently "authentic" than any other.

PUBLIC SCREEN/WINDOW/VISION

Several artists and writers working during the late 1970s and early 1980s—as well as later historians taking stock of politically engaged, emancipatory projects and practices of the time—similarly looked to methods that, in contrast to predominant voices of US postmodern discourse, allowed for experiences of fragmentation and alienation to yield new imaginings of agency and participation in cultural and social production. In particular, certain feminist and activist artists sought to create theoretical and analytic platforms for potential resistance to the new discursive hegemonies of "spectacle" and "simulation," a resistance based on the ambiguities, inconsistencies, and heterogeneities of encounters between images and perceptions. Some, such as Rosalyn Deutsche and Martha Rosler, in their reassessments of urban space and new media as crucial sites of struggle over public engagement, turned to Henri Lefebvre's "right to the city," Jane Jacobs's "uses of the sidewalk: safety and contact," and Oskar Negt and Alexander Kluge's constellation of plural "counterpublic spheres."[60] Griselda Pollock and the editors of the influential film and media journal *Screen* championed the work of Bertolt Brecht as a way to politicize the traditionally binary structure of public and private as exclusive spheres of experience, a method that would account for a kind of seeing and being not easily compatible with existing formats of homogeneity and difference.[61] In her 1982 essay, "Whose Brecht? Memories for the Eighties," Sylvia Harvey pondered whether historical avant-garde models such as Brecht's, in their reliance on ostensibly retrograde ideas of socialism, scientific progress, and the possibility of an emancipatory popular culture, could serve as a much-needed creative reinvigoration of political imaginaries in a climate of intellectual pessimism and reactionary populism.[62] And the writings of Brecht's colleague and friend Walter Benjamin rose to great prominence with their optimistic promises of historical redemption, brought about by the very mechanisms of mass production and consumption that to so many others signaled the inescapable demise of the modern enlightenment project. Both Benjamin's "The Work of Art in the Age of Its Mechanical Reproducibility" essay and the *Passagen-Werk* left a lasting impression on the art and media discourses in the United States. Published in translation as early as 1969 (the "Work of Art" essay) and discussed by scholars including Susan Buck-Morss and Miriam Hansen in publications ranging from *New German Critique* to *October*, these texts offered nothing short of a "theory of experience," or rather a theory of empowerment through dialectical perception, through what Michael Jennings and Tobias Wilke call a Benjaminian constellation of "reception, perception, tactics, and aesthetics."[63]

Particularly compelling for the context of *Hole in Space* is Benjamin's connection between the (self-)perception of the urban subject and the cinematic screen, both of which offer critical and interrelated modes of modern sense-abilities that transcend conventional narratives of mass culture to avail an organization of experience in emancipatory fashion. As cinematic displays of performed sociabilities on view in the windows of cultural and commercial nodes of consumption, the screens of *Hole in Space* can be understood in terms of the "self-alienating" mechanisms Benjamin saw at work both in the flânerie of the modern metropolis and the technological apparatuses of reality's reproduction. After all, Benjamin himself thought of the "Work of Art" essay as a "telescope" that would help him look through the "bloody fog" of the "phantasmagoria of the nineteenth century" in order to, as Hansen puts it, "delineate in it the features of a future, liberated world."[64] Mobile Image similarly saw (and used) technologies of representation and perception as tools for the repositioning of the social subject as a multiplicity of past and future potentials, oppressive as well as liberating. Ultimately, what *Hole in Space* achieved for three days in 1980 was less the technological resurrection of a romantic ideal of urban public life—replete with the joys of unmediated human interaction, civic intimacy, and thus symbolic liberation from the ideological and material constraints of rapid techno-economic expansion and neoliberal individuation—than a self-conscious performance of the remnants of historically constructed desires for authentic subjectivity available primarily as images and objects to be consumed. (Galloway had had first-hand experience with this process of constructed desires and their satisfaction through commodities, having been hired by the Sony Corporation in the early 1970s to design interactive multimedia installations for the company's Paris storefront on the Champs-Élysées.) On display in the (shopping) windows of Lincoln Center and the Broadway Department Store were fantasies of identity and belonging through civic rites of spontaneous communication and interpersonal exchange as already scuffed and outdated products. Some participants recognized this nuance amid the work's celebratory atmosphere. "It could have been done thirty years ago," one woman reacted, "Why did it take so long?" *Hole in Space* was an exercise in transforming, to use Benjamin's words, "dream images of the collective" into "historical images," thus laying the groundwork for Mobile Image's future project of creating counterpublic spheres of self-organized experience.[65] Crucial to the historical assessment of this work is how it both framed and enacted "the city," "technology," and "the public" as a set of given and potential parameters.

In her 1983 essay, "Benjamin's *Passagen-Werk*: Redeeming Mass Culture for the Revolution," Buck-Morss presents the German philosopher's interest in the modern city and industrial techniques of pictorial production as an attempt not to oppose but rather to utilize and refunction the objects of mass consumption. To Benjamin, Buck-Morss argues, the commodities featured in the shopwindows and arcades were accumulations of past promises and lingering desires, "a world of secret affinities."[66] These objects were collective dream images, reifications of an entire culture's subjectivity rooted in ideals of innovation, affluence, freedom, and control. As commodity fetishes, they bore both the lack of real-life materialization of those ideals and the insufficient and ephemeral gratification of these ideals *as* objects, whose function was fully grasped through neither their exchange nor their assigned use value: "Benjamin affirmed the mythic power of wish-images which found unconscious, symbolic form in commodities and mass culture. But as dream-images they were fetishes, alienated from the dreamers, and dominating them as an external force."[67] This phantasmagoric object culture expanded beyond the shopping windows to encompass the entire city—"buildings, boulevards, all sorts of commodities from tour books to toilet articles"—a reality-as-artifice made possible by new industrial processes.[68] And looking to Marx's famous observation regarding the commodification of social relations, Benjamin affirmed the transformation of affective and communicative connections between people under capital "as what they really are, material relations between persons and social relations between things."[69] This is how the flaneur perceives others and ultimately himself in the panoramic city—at a distance, at home in the crowd, strolling, browsing among objects, a distracted and disinterested gaze providing an experience of orientation and control. Writing in 1983, Rosalyn Deutsche proposes a similar (re)reading of Ludwig Kirchner's early twentieth century images of metropolitan life.[70] Deutsche takes umbrage with the canonical account of the artist's distorted and grotesque urban figures as representing a subjective and thus genuinely existentialist-humanist reaction against the cold and isolating forces of the modern condition. Rather than relegating Kirchner's work to a symbolic, timeless defense of the human spirit against a mythical and naturalized technocracy, the author seeks to politicize paintings fashioned on the eve of a social and economic crisis: "Kirchner went beyond a mere depiction of alienation to observe its actual cause—the dominance of a money economy."[71] The works present their audience not with the corruption of an essential humanism but rather with a historically determined form of being, "modeled on commercial transactions and centered on acts of display, buying and selling": "Contemplating the city crowd, Kirchner depicted its members as commodities . . . [he] focused on the objectification of human relations inherent in economic exchange."[72] Available for consumption on the street, in the store windows, and in the paintings themselves are the artificial images of modern togetherness, of the exciting and absorbing bustle of urban public life.

In many ways, *Hole in Space* functioned as precisely this type of display of objectified ideals—here, not just depicted but also enacted on the spot. What the participants encountered in the windows, on the screens, were images of spontaneous, unfiltered interaction, of placemaking and subjectification through a real-time connection between here and there, a simultaneous transcendence and affirmation of rootedness in space

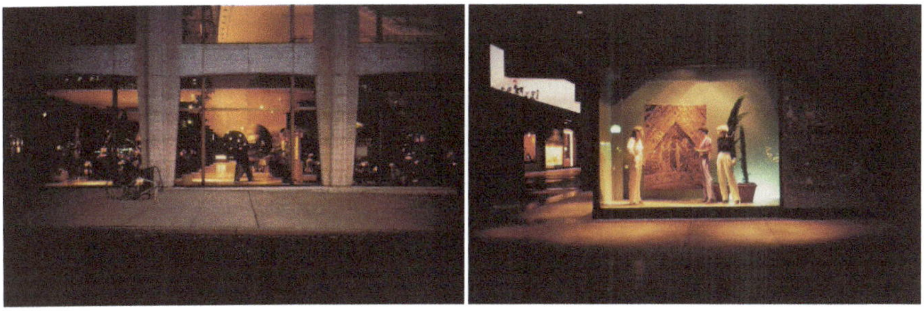

through the observation of others as a collective body, always on the edge of forming and dissolving. People continuously walked in and out of the picture, and the cultural sense of belonging via a playful standoff between East Coast and West Coast attitudes was ever tenuous. On those screens—one in the window outside of Lincoln Center, a cultural-economic node that has drastically contributed to New York City's violent reconstruction of urban space and its population, and the other in the window of the Broadway department store in Los Angeles, an unabashedly private space of communal consumption in a city that is said to be all spectacle (figure 2.6)—unsuspecting passersby caught a glimpse at the joy and awkwardness of what late capitalism sells as genuine sociality. By placing that sociality in a shopwindow, Mobile Image underscored its status as commodity, as something on display and to be readily purchased; *Hole in Space* quite literally depicted "the objectification of human relations inherent in economic exchange." As Deutsche explains of Kirchner's images of prostitutes, "Their symbolic containment within the store window's reflected glow effectively situates them as objects for consumption," rendering them emblematic of modern urban life, in which "behavior is always

2.6

Locations for *Hole in Space (A Public Communication Sculpture)*. Left: Broadway Mall, Century City; right: Lincoln Center, New York, c.1980. Courtesy Rabinowitz/Galloway Archive.

modeled on commercial transactions and centered on acts of display, buying and selling."[73] Yet, unlike modernist models of "disinterested" vision, in which the viewer occupies an idealized space, removed from circumstances of production and reception and from which he or she can see an image in its self-contained totality,[74] the participants in *Hole in Space* could not remain solely detached contemplators of symbols—although nor, conversely, could they wholly lose themselves in the act of cross-country communication. The work placed viewers simultaneously in front of and behind the camera (and window)—rendering them both performers and audience, consumers and consumed. Affixed to the very spaces designed to peddle taste, a sense of ownership, of choice and ability and control, thus "interpellating," to use Louis Althusser's term, the subject *as* subject, state-of-the-art communications technology offered its viewers at least the potential to recognize the alienating experience of their very own participation in the processes of fetishization and consumption. With the apparatus hidden from view—cables and cameras, operators and sponsors all tucked away, off-stage—the experience remained with the encounter as such, the interaction with a moving image, itself based on ideas, expectations, memories, and desires.

As mentioned in this book's introduction, Tom McDonough denounced *Hole in Space* in his review of the 2008–2009 exhibition "The Art of Participation: 1950 to Now" at the San Francisco Museum of Modern Art. Basing his critique on the exhibition's documentary installation, almost thirty years after the fact, in which viewers were passively pinned between two screens on which the originally recorded exchanges were projected, McDonough sees the work as complicit in a culture of ego gratification and postmodern self-recognition through techno-bedazzled exteriorization. Yet, this outward projection, in *Hole in Space* as well as through other new-media tools, cannot simply be understood as "a reassuring encounter with the self."[75] There indeed may be no such thing as an inherently authentic self in the individualist-modern, bourgeois sense, but neither has that self been dissolved into pure artifice and imagery. As a fundamentally Benjaminian project, *Hole in Space* portrayed and performed subjectivity as concept, as site, as a materialization of expectations and ideas, as an experience in process. The integrity of art and the artifice of commerce that McDonough posits as the markers of a modern culture in decline were in fact articulated as sides of the same coin, the shopwindow of Lincoln Center and the Broadway of the mall. Mobile Image was not concerned with the demystification of artmaking but rather with the politicization of the technologies of aesthetics, the sites and tools of social construction.

It is tempting to read *Hole in Space* as a temporary fix to broken urban space, as an attempt to reconstruct the public sphere through electronic connectivity in places either occupied by private interests or which function as mere passageways for hastened traversal and aimless meandering in between what Castells calls nodes, at a time when, as he puts it, "public space is being reduced to the space of leisure . . . [and] wandering."[76] But, as Castells reminds us, new communication services tend to exacerbate spatial inequality: "Technology assists to a new step toward the disintegration of urban cultures that were always characterized by the mixture of uses. Land use zoning started to segregate people and activities. Now 'electronic zoning' dramatically enhances this tendency, transforming places into unifunctional units."[77] Even though *Hole in Space* resembled and conjured the commercial and strategic surfaces of information exchange and entertainment, it did not fully coincide with them. The work's screens were temporary and evolving, adjacent to the citadel of culture and the temple of consumption, as if to provide glimpses at what is essentially being offered on the inside, windows onto a world of ephemeral belonging through others and/as objects. Mike Featherstone asks whether flânerie as critical method, as the self-aware mode of reading the texture of urban life, can find its correlate in "virtual public life."[78] Does the "city square-cum-screen" offer even greater freedom of movement, a disembodied gaze liberated from the constraints of corporeality? Yet, to a crucial extent, public life has always been virtual, based on the media as purveyors of communicative action as much as on the sites where individuals could congregate in person. Habermas defined "publicity" (*Publizität*), the mediation of information and opinions through "daily political newspapers" and "literary journalism," as a—maybe even *the*—central component of the historical public sphere (a role later extended to "newspapers and magazines, radio and television" and compromised by an "influx of private interests" and monopolization). During the nineteenth century, he explains, "the appearance of a political newspaper meant joining the struggle for freedom and public opinion, and thus for the public sphere as a principle."[79]

Hole in Space acquires a more useful function when seen as an investigation of an ideology of progress that substitutes novel technological devices for social innovation, raising questions regarding the quality of access to and utility of the sites and tools of information and communication. The public's medium (or mediation) does not shift from city to screen; both change in their relationship to one another as material and ideological entities. Writing about "video in the public sphere," Martha Gever insists that "the medium, of course, is television. But not *television*."[80] The apparatus facilitates ways of seeing, but it is not them. In a similar sense, the window screen of *Hole in Space* engages the city as public, as image, as panorama, as spectacle, but it doesn't take its place. In Benjamin's account, according to Tobias Wilke, "the medium names the comprehensive force field that links human sensorium to world and that is constituted in doing so by the interplay between natural (physiological, physical) and historical (social, technological, and aesthetic) factors."[81] *Hole in Space* links an experience of what it means to be public and to be in public in 1980, in the urban United States, to the sites and tools of its production and performance, articulating an aesthetic politics of commercial and cultural re-presentation, identification, and self-actualization. As for Benjamin, the issue of *how* this experience is organized remains. Not in all formations or under all circumstances does the "force field" become tangible, intelligible, "an analytical tool [serving] to delineate and render accessible to investigation a *dynamic process of 'organization'* in which different determinants interact to generate historically specific economies of perception" and, ultimately, of transformation.[82] According to Hansen, Benjamin's "Work of Art" essay shifts "the emphasis from a definition of what film [and photography] *is* to its failed opportunities and unrealized potentials."[83] The media under discussion are not automatically emancipatory but need to be used as such.

Certain artists, Mobile Image among them, thus attempted to refunction or rather unleash "mass" culture's liberational potential. Along with some of the aforementioned artists, in the 1980s, Krzysztof Wodiczko, Alfredo Jaar, Jenny Holzer, and others not only infiltrated urban space in ways that reframed its definition, utility, and meaning but also, like Mobile Image, explicitly took up the commercial technologies of publicity—billboards, posters, store windows, facades, marquees, and screens of various kinds—as sites of counterpublic intervention. The collective dream images peddled by artistic and commercial artifacts alike would therefore be seen as historical images, as containers of exhibition value out to provide just enough symbolic gratification for many a genuine desire. In his series of public projections, for example, Wodiczko used state-of-the-art cinematic technology to juxtapose the aesthetic and socioeconomic dynamics of the struggle over so-called urban revitalization, casting images connoting homelessness, worker exploitation, and the marginalization of those "social elements" deemed a blight for newly refurbished public spaces and the social harmony they symbolize onto civic buildings and monuments, architectural reminders and eternally embodied exemplars of the deeds of great citizens and national heroes. The artist considered this an attempt to produce a new form of being in public, where ideas and forces clash, where power becomes visible instead of hidden and therefore subject to contestation, where who and what is public is a matter of evolving historical parameters.[84] What Wodiczko achieves through his work, as Rosalyn Deutsche argues, is not a "better" or "more complete" image of a particular urban space but rather a multiplicity of voices and competing interests and perspectives conveniently paved over in the name of cultural stability.[85] Holzer's *Truisms* likewise partook in the ambiguities driving the battle over site and sights, treating public space as a place of inscription, of visible and invisible production of knowledge and meaning. The series engaged the public sphere as a variety of competing sites.[86]

It also examined some of the usually unexamined myths broadcast in public spaces via communication technologies, adding rather than simply substituting different "truths" by presenting different perspectives and voices, thus expanding what Negt and Kluge call the "social horizon of experience," opening it up to inquiry, multiplicity, and contestation. This expansion is not solely that articulated by Habermas and other theorists, who see such contestation as effectively blurring the lines between those spheres traditionally kept separate so as to guarantee the public's idealized autonomy. It is a struggle over how what is deemed to be in and of the public interest represents and serves only a select, exclusive part of the social body, while determining to a great extent, and in many cases rather forcefully, the experience of all. In this sense, Craig Owens is right to criticize critics such as Benjamin Buchloh, who, in their albeit thoughtful and thorough analysis of feminist art, neglect to think beyond the manipulative function of mass culture, to consider the demands made by such art as more than an exercise in deconstruction. In this type of discourse, according to Owens, artists such as Holzer, Dara Birnbaum, Louise Lawler, Sherrie Levine, Martha Rosler, and Barbara Kruger "are said to manipulate the languages of popular culture—television, advertising, photography—in such a way that 'their ideological functions and effects become transparent'; . . . in their work, 'the minute and seemingly inextricable interaction of behavior and ideology' supposedly becomes an 'observable pattern.' But what does it mean to claim that these artists render the invisible visible, especially in a culture in which visibility is always on the side of the male, invisibility on the side of the female?"[87] Holzer's work "unveils" a so-called female point of view and the truth that the public subject, the default body and voice, the image of individual and social selfhood promoted by commercial and cultural apparatuses alike, is gendered, and, by extension, determined by sexuality, race, and class—all of which are structurally relegated to the so-called private realm. What is added here are experiences of and perspectives on fatherhood, patriarchy, and romance that are part of the experience determined by the dynamic of social ideals and performance but which appear incommensurable with the imagery serving to sustain that dynamic.

Seen in this context, *Hole in Space* offered a similar montage of competing sensibilities, of expectations and enactments. Although it certainly seemed less overtly "political" in its messaging, it became a comparable tool of observation and politicization through self-alienation—a critical if not necessarily always immediately conscious distancing through the identification and dis-identification of the performers from the roles on display. The bonds among those gathered on either side of the techno-bridged void were reduced to the most trivial commonplaces (e.g., sunny West Coast laissez-faire vs. New York tough-guy attitudes) and benign interactions (e.g., a game of charades), while individual performances for the most part reproduce stereotypically gendered (and, to a lesser degree, ethnic/racial) representations of bravado and hysteria, seduction, and vulnerability. Pamela Lifton-Zoline witnessed an exchange, for example, that seemed to her to be extraordinary in its conventionality, in its enactment of a "type" before a cheering crowd:

RITA: Hey, you're back, what'd you bring us, I love you [grin] man, what's your name?

AL: [Holding up bottle of Heincken] Al.

RITA: Hi, Al.

AL: [Holding up bottle in a toast] Where you goin'?

RITA: I'm standing right here talking to you.

[in NY Al shimmys [sic], and then, in LA an answering shimmy]

. . .

AL: Hey girl, we got to get together some time.

RITA: I'll come by and look for you <u>personally</u>. [cheers]

. . .

AL: You got it, you got it baby, you got it; my number is 622-9796 woo woo woo.

RITA: Area code?

AL: 212, baby.

[A pause in the conversation while other people talk back and forth]

RITA: She [my friend] just called [your number] and <u>some lady</u> just answered. Who is that?

AL: I'm not married [there follows some kind of explanation of people staying at his place, etc.]

[Talk of airline fares, etc. and of climate]

AL: That's ok, I'm sure you'll provide enough heat for me.

. . .

RITA: Do you do cocaine?

AL: Do you do it? [laughter, giggles from both sides]

RITA: My middle name's cocaine.

[Rita and Al confirming plans for phone call, front-pocket/back-pocket sex joke, finish]

In *Hole in Space*, identities were both deeply inscribed and worn thin, the unease of encounter—in both its rudimentary humanism as well as the attempt to relay something more intellectually profound—made palpable. At its most successful, the work went beyond modernist deconstruction to what Owens describes as "seeing difference," not in the sense of a pluralist postmodernism but rather as the politicized understanding of difference as constructed relationality, as producing generative indeterminacy rather than adhering to absolute categories.[88] A counterpublic exists not where traditional publicity is expanded through novel means and multiple voices but rather where the very structure of the public sphere as social horizon of experience is challenged in its coherence through the exposure of its self-sustaining mechanisms of differentiation. Benjamin describes the flaneur as maintaining a safe distance from the crowd, an ability to retain a bourgeois, patriarchal sense of selfhood, precisely because he did not see the urban masses in terms of class.[89] Urban publicity is only truly emancipatory when ideological and technological means of identity ownership are refunctioned. The panorama, cinema, and new media have extended the "fluidity of flânerie " to anyone. Yet, as Featherstone cautions, despite this imaginary "mobility," "women and ethnic and racial minorities rudely discover the bodily truth in the differential between this virtual position and the real."[90] Hegemonic bodies and attitudes—individual and social—dominate the stage and the screen, and the techno-enabled peruser of the urban encounter runs the risk of falling for the same facades Benjamin outlined decades prior: "In fact, this collective is nothing but appearance. This 'crowd,' in which the flaneur takes delight, is just the empty mold with which, seventy years later, the *Volksgemeinschaft* [People's Community] was cast."[91] To Benjamin, a depoliticized collectivity—that is, collectivity as such, without an emancipatory, progressive sense of social utility—is an easy target for fascist appropriation.

In his discussion of video art, Buchloh raises a similar concern with regard to art and technology's interpellation of "the public." In particular, he focuses on Holzer's *Sign on a Truck* (1984; figure 2.7), which is arguably the work that is most similar to *Hole in Space*. Holzer's work consisted of a mobile thirty-foot video screen, akin to those used in sport stadiums, installed on two different days in two different midtown and downtown Manhattan locations, just prior to election day. Meant to expand, or at

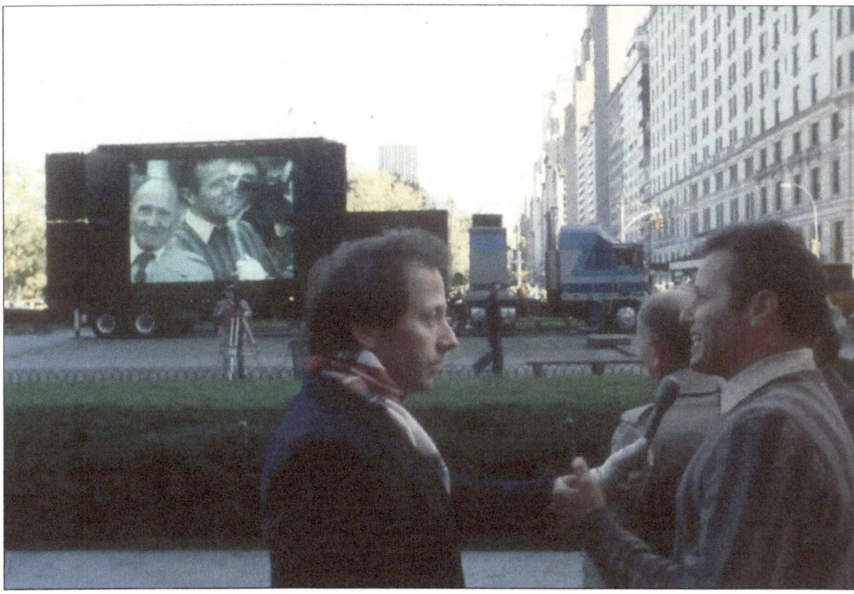

least test, the idea of new media's popular access and impact, the work displayed prerecorded images and messages by several artists and authors as well as interviews conducted by Holzer and her collaborators with passersby in the street. People were asked about their concerns and opinions in the days leading up to the elections. Furthermore, open-mic sessions provided the opportunity for what Buchloh calls "the direct interference and participation of the viewers in the process of forming a visual and verbal representation of the political reality of the viewers."[92] And although he acknowledges the project's seemingly successful continuation of the Russian avant-garde's deployment of mobile consciousness-raising devices in the form of agitprop trains, trucks, and boats, as well as the integration of contemporary technology "with the needs of the late-capitalist urban public and its peculiar form of illiteracy," Buchloh

2.7

Jenny Holzer, *Sign on a Truck*, 1981.

finds one of the work's most valuable contributions to be its revelation of "considerable problems," namely that it reinforces the myth of "the collective mind as being innately democratic . . . a myth that itself functions to protect us from insight into the actual operations to which the collective mind is subjected."[93] In order to prevent a mere regurgitation of imposed ideologies, a project like *Sign on a Truck* needed to rupture such a myth, making visible the constructedness of reality and countering the simplistic belief in inherently progressive qualities of technology and "authentic" sociality, especially when reproduced in idealized situations of artistic experimentation.

What is missing from this astute concern, however, is a more nuanced notion of the public sphere as a site of struggle. Whereas Buchloh points to a "liberal ambivalence" when, as part of what he deems an intentionally democratic-progressive project, it turned out that many people interviewed for and during the work were fervent Reagan supporters, he does not see the displays of opinion and attitude, the articulations of perspectives, fears, and desires, as anything but the expression of a superimposed and passively, unambiguously consumed false consciousness. Which roles and types are affirmed? Which ones thwart expectations, if ever so slightly, with regard to who is assumed to express what opinions, outlooks, and needs? Does the sum of utterances add up to a complete picture of what democracy and public opinion ought to look like, on the street, at these locations, on a giant TV/entertainment/advertising screen? Or is there a form of montage, of defamiliarization available as the artist juxtaposes artist-created clips and two-party political opinions along with a variety of faces, of points of view, and of reception, on and off screen, some seemingly in and others out of place in the mediated visual and physical sites of urban capitalist culture?

MEDIATION

As a function and mediator of norms and possibilities of publicity, connectivity, encounter, and positionality, *Hole in Space* resonates in a range of more recent discourses. It relates, for instance, to Erkki Huhtamo's demand for a "media archeology," while its complication and articulation of the subject position of the motile urban self relates to ongoing developments in critical urbanism.[94] In fact, in line with Castells's and Sassen's work, recent scholarship in both fields has argued that the two—media and the city—need to be considered in their historical and continuous intertwinement, as codependent entities and as apparatuses and technologies that organize and manage the dynamics of contemporary existence. Addressing the complex and often disjunctive interplay between images, imaginary, and experience, Huhtamo points out that "visual culture does not exist solely in interior spaces" (and, by extension, neither does the private). Scott McQuire argues that the current state of the city as "media-architecture complex" has been underway "at least since the development of technological images in the context of urban 'modernization' in the mid-nineteenth century," with its full effects and implications materializing with the extensions of digital networks.[95] Andreas Broeckmann posits the urban itself as media/medium: "The 'screen' of the city is a cluttered kaleidoscope . . . a cacophony of mediated messages," he writes. "The city is a field of experience in which built architecture, visual displays, and personal communication overlap to form a rich, often overly rich mesh of impressions."[96] Sassen discusses the inadequacies of "understanding a city through its built topography" in a digital age, offering instead a model in which a given city is but one site in a global network of "circuits" through which a range of operations—economic, political, cultural, subjective—are distributed.[97] (As will be discussed in chapter 3, this very phenomenon will become more central with *Electronic Café*, which in many ways reflected and reflected upon Los Angeles's concurrent development into a "world city.") Arjun Appadurai theorizes the "new global cultural economy" under "disorganized capitalism," a complex no longer sufficiently charted by traditional geographies (e.g., nation-states or urban–rural dichotomies) or models (e.g., "center-periphery," "surplus and deficits," "consumers and producers"). He proposes a framework of overlapping and layered "scapes"—"enthoscapes, mediascapes, technoscapes, financescapes, and ideoscapes."[98] In their dynamic interrelatedness, these "scapes" serve to map existing worlds (images) as well as possible ones (imaginaries) in all of their overlapping and disjunctive complexities, technologies at once descriptive and generative.

Possibly the most ambitious attempt to grasp and find agency within the "new normal" that is the (still?) current state of such connections, territories, and networks is Benjamin Bratton's theory of the Stack.[99] Bratton aims to move beyond the notion of a "hybrid urbanism"—the idea that "physical/virtual mixtures of bricks and bits are still a mysterious novelty"—and instead come to terms with the fact that "the new normal twists distant sites into one another": "Discontiguous megastructures cohere from molecular, urban, and atmospheric scales into de facto jurisdictions. Ecological flows become a public body of intensive sensing, quantification, and governance. Cloud platforms take on the traditional role of states, as states evolve into cloud platforms. Cities link into vast and tangling urban networks as they multiply borders into enclaves inside of enclaves, nesting gated communities inside of gated communities. Interfaces present vibrant augmentations of reality, now sorted as address, interface, and user."[100] The Stack consists of several interwoven and interdependent platforms: "*Earth, Cloud, City, Address, Interface, User*." It is material and immaterial, both hardware and software, an "accidental megastructure," a new logic of "vertically thickened political geography." It is, Bratton explains, "a machine that serves as a schema, as much as it is a schema of machines."[101] As a totalizing theory, the Stack is strategic. It "make[s] the composition of new governmentalities and new sovereignties both more legible and more effective," offering insights into the constantly changing and thus arguably change-able constructedness of a shared contemporeality.[102] It can also present important retrospective questions about a work such as *Hole in Space*. If *Hole in Space* is part of a much-needed history or archeology of urban media (and the urban as media), then what does Bratton's lens of contemporary technological interrelationalities, of layers/scapes/platforms, tell us about that work's function regarding the intertwined scales and modalities of screens, sites, subjects, and mobility at the time? What lessons does it hold today? And what effect does the aforementioned contestation over ownership space and tools have in such a model—that is, how does what Anna Everett calls the gentrification and "domestication" of communication technologies affect the User and the Interface in the platform's logic and manifest dynamics of power and access?[103] In the Stack, what does it mean to, as Everett puts it, "Take IT to the Streets"?[104]

The "right to the city" then means "the right to the Stack." Like Mobile Image and figures such as Enzensberger and Youngblood before them, new media historians, theorists, and activists struggle to approach technological "governmentalities and sovereignties" as a dialectics of subjugation and emancipation. As Huhtamo explains, precursors to today's mobilization of publics through technical interfaces were said to immerse the spectator in a process of perceptual, spatial, and social alienation and intimacy, a "bipolar optics."[105] Magic lantern shows, popular in the United States during the second half of the nineteenth century, consisted of slides projected outdoors on screens, walls, and shopwindows featuring the "latest news bulletins or election results"; some saw these events as having been lively spectacles, while to others, they reinforced the passivity of the onlooker.[106] Both Huhtamo and McQuire point out that the development of large urban screens (panoramas, roof projections, billboards, LED media facades, etc.) have, since their inception, been accompanied by corresponding miniature versions, ranging from desktop "peepshows" and "moving panorama handscreens" all the way to today's mobile devices.[107] The advent of these technologies is part of the same paradigm shift concerning spatial mobility, creating new patterns and perceptions of movement across what had been presumably stable positions, whether topographical, social, or psychological. Studies of relational space and "sociality at a distance" in mediated urbanity have tended, according to McQuire, to focus on either spectacle or surveillance.[108] As telecommunication technologies present new configurations of mobility and traversal, the potential for the emergence of relationships beyond already existing and circumscribed social formations—those that rely on physical face-to-face and face-to-object encounters in localized sites—is hindered by the increasing individuation and "rampant individualism" born of alienation, disaffection, and the reduction of agency to choices of consumption and ostensibly tailor-made experiences. Concurrently, as surveillance is increasingly internalized, the same tools that open up new ways to transmit and travel enable new forms of control. That internalization stems less out of fear of being watched than through the assertion of one's presence in an exceedingly complex and multilayered landscape of everyday existence. McQuire thus asks: "Is there still space for social interactions outside the dictates of surveillance and the spectacular forms of commodity display?" Seeking to move beyond the "self-fulfilling prophecy" of denying viable alternatives, he poses the question not of if but of *how* such interactions can be facilitated.[109] McQuire subsequently looks to Richard Sennett and Alexander Kluge, proposing a model based on "play," "ritual," and "embodied participation." In many ways, this recalls the efforts of Mobile Image with *Hole in Space*. In the example of Rafael Lozano-Hemmer's *Body Movies* (2001–2006) and *Underscan* (2005–2008), digital projection and shadow play invited passersby in urban public spaces to interact with pictures and portraits, producing a "complex interplay of body and image, . . . of personal intimacy and civility."[110] The potential, McQuire argues, lies in the created awareness and practice, as participants discover that they can intervene in the urban environment, "enabling the city to become an experimental public space."[111] If, as Wendy Chun argues, "community is communication and space," Mobile Image's invocation of an urbanism based on a playful, creative, and collective contestation of publicity and the publicness of bodies and images remains a viable model for progress today.[112]

Although many of these texts and studies acknowledge the problem of the uneven access to tools and networks and thus the dangers of recreating an idealized version of the public sphere based on fabled notions of inclusivity, few discuss how the mechanisms of proliferating new media actively undermine the very agency they ostensibly set out to provide. Critical urban theory and practice should make visible and tangible both the pathways and the partitions of the mediated and interconnected landscapes and articulate their constructedness, causes, and functions. It would have to show and put into play how screens and interfaces are embedded in the city fabric and how the subject is embedded in the city and in the screen. In his examination of metropolitan publics, Broeckmann calls for technical media to create a form of intimacy that is not reducible to consumption, entertainment, and information, but rather becomes one of several sensory interfaces that we use to look, listen, touch, smell and taste.[113] How, then, can we posit and use the screen as a way to engage outward and inward from a subject position defined by a set of relational mechanisms that connect (and transcend) (our)selves on an embodied and biopolitical level? Broeckmann also considers the urban screen as facade, decoration, and camouflage, as a way to hide, cover up, or simply redirect one's sensibilities (and sense-abilities) from the traces and presence that make the network of metro-flows and metro-economics stutter—whether that be graffiti, poverty, or social injustices. As such, urban screens are part of the aforementioned history of technological disembodiment and deracialization traced by Lisa Nakamura. They contribute to a politics of placemaking and Deutsche's "uneven development" in a topographical-architectural sense, as much as they perpetuate the myth of universal participation that de-emphasizes the segregation of physical space in favor of the supposed technical erasure of material boundaries.

In *Digital Diaspora: A Race for Cyberspace*, Everett discusses the codependent evolution of postwar spatial and telecommunicational reordering in the United States. As struggles for racial and gender equality demanded an increasing presence on urban streets and television newscasts and courts "mandated forced racial integration of American uncivil society," she explains, the "erosion of public confidence in the safety of the nation's public spaces" was (and continues to be) countered through white flight and gentrification.[114] At the same time, the installation of TV sets in American homes created a "highly constructed 'antiseptic' image of social space," one rooted in the utopian discourses around earlier communications technologies (e.g., telephone and telegraph), promoted as purifying agents. Such discourses helped normalize racial and spatial divisions for the postwar suburban middle-class audience. "Not surprisingly," Everett argues, "this 'antiseptic model' of mechanically and electronically driven participatory democracy has morphed into present-day utopian discourses promulgating the new digital democracy as society's panacea for the dawning millennium."[115]

Any suggestion that technologically mediated and amplified space has the potential to reverse what Jonathan Crary and others have charted as the creation of modern spectatorship—in which the carnivalesque participation in culture and public life of the unruly crowd gave way to the increasingly individuated and privatized consumption of images and perspectives—has to consider this doubly disembodied cleansing.[116] If the term "carnivalesque" is to be strategically employed, as Huhtamo and McQuire do when they find in public screens the potential for an "ad hoc carnival," it must be posited in a new way: it must refer to the raucous (dis)organization of assembled subjects and to the reconceptualization and animation of selfhood as leaky and embedded, multiple and permeable, connected and contested.[117] Rather than transcend bodily physicality, "new kinship structures," as Everett calls them, have to connect embodied subjects according to a new logic.[118]

This new logic is, to some extent, modeled by *Hole in Space*—and theorized more recently by Bratton. The Stack prevents us from thinking the various platforms, levels, and layers apart, from perceiving them as cleanly separate, and thus makes it impossible to reinscribe the same logic of public access to technology as was formerly done to space. Of course, the Stack exceeds the multiplicity of *Hole in Space*'s modalities, mediations, and encounters, but the work pointedly corresponds to three of its key platforms—the urban ("City"), the screen ("Interface"), and the subject ("User")—all of which connect and entail parts of the other. "The Stack," Bratton explains, "is not only mediated through the City layer; the entire apparatus also expresses itself at the scale of the City and the built environment."[119] *Hole in Space* is part of the Stack and, to some extent, *is* the Stack, or at least part of its genealogy, its archeology. If the work's main contribution to the history of art and technology is its critical and reflective positioning of the urban self/User, the question is what kind of User is and can be constructed through the work, in the Stack, in ways that do not repeat the depoliticizing and disembodying perpetual reproduction of individual and social subjecthood as ostensibly discrete, autonomous, and universal? Or, as Bratton puts it, how can one foreground "the human mind, body, and species" as "an open field irreducibly dependent on forces both larger and smaller than its own shell" and "erode conceptual models of the atomic individual as historical actor"?[120] Rather than affirming "the individual [who] sees his own poetic reflection everywhere," does Mobile Image manage to stage subjectivity as an "always manifest image cobbled in relation to available technologies of self-reflection"?[121]

Over the course of the three days, the work explored numerous permutations of technologically facilitated publicity in urban space. At its most successful, it did not merely reiterate and reproduce the mythical trappings of public encounter and techno-social progress and the types of prescribed individual and social subjectivities they ostensible avail but also produced critical perspectives onto such notions through conflicting

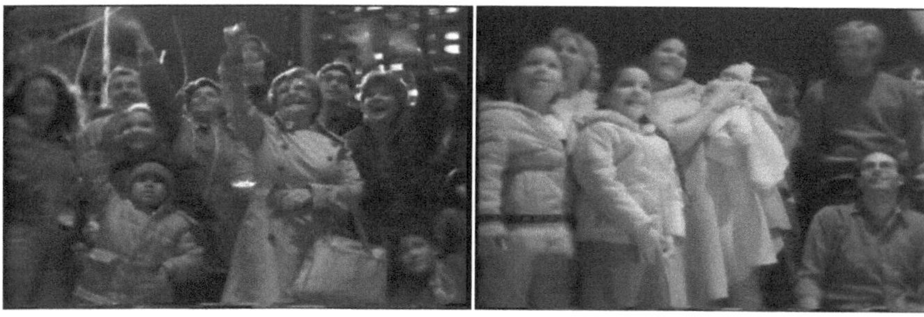

experiences based on the ambiguity and incompatibility of ideal and real, projected and lived sociality and selfhood. Yet, this was certainly not always the case. Some participants were unselfconsciously swept up in the apparent spectacle and euphoria of seemingly unfettered access and connection (figure 2.8). This was somewhat disappointing to the artists, who noticed that *Hole in Space* favored a narrow set of behaviors and personalities, and that many failed to think beyond their own constructed longings and thus fully consider the implications of the work. Some were brought to tears simply by the satisfaction of seeing a relative or old friend for a few minutes; others speculated about potential uses of the technology, but in utterly conventional terms, such as when two men discussed their desires to use it for more efficient business meetings or to enlarge the reach of information distribution dramatically. For others, the work served as a negative critique, frustrating those who only found their alienation from urban and mediatized space confirmed, or who felt the sting of reification and commodification of social interaction amplified through the project's temporal, ephemeral gratification of the desire for connectivity, exchange, and intimacy. As mentioned, some recognized that the technological

2.8

Mobile Image, *Hole in Space (A Public Communication Sculpture)*, screenshots from video documentation, 1980. Courtesy Rabinowitz/Galloway Archive.

capability on display was not new and wondered why these devices were not already readily available. Another participant immediately recognized what was excluded from the experience—the people and places in between the two most populous US cities—wondering "can we change the channel to Pittsburgh?" Maybe the montage that was *Hole in Space* came into view only over the duration of the work, as a techno-sociological experiment on unwitting subjects, the work functioning as the microscope Galloway imagined, an outlook onto and model for the subsequent *Electronic Café*. This next step was made possible, even necessary, because *Hole in Space* never allowed for a complete assessment, an analytical totality, whether on the ground within the actual piece or in its immediate aftermath.

Acting in and for the screen as part of a changing assembly of overlapping and interrelated identities, all participants in *Hole in Space* became both subjects and objects. As discussed, being part of the crowd in one's physical location and to those to whom one appeared on the screen produced an often awkward but reflective self-consciousness, reinforcing division even as it brought people together. Viewers struggled to connect on both sides, individually and collectively, latching onto geographic-cultural particularities, presenting personas, performing sometimes internalized, sometimes overtly mediatized roles—pick-up artist, policeman, infatuated lover, businessman, doting mother—with lines blurring between the two. Both literally and ideologically, *Hole in Space* precluded a centered, detached position from which to unify the work, or oneself within it. At times, the joy of connecting was uncomfortably theatrical and artificial; at others, the theatrical was the most pleasurable part. On the second day, once word had spread, people arrived with the intent to perform for the camera, to broadcast themselves, some as recurring, recognizable characters throughout the brief life of the work, others to promote and advertise their talent or enterprise, such as a kid playing the violin on his mother's behest, or the cast of *Evita* announcing the run of their show on Broadway. Whereas such behaviors may have diverged from Mobile Image's initial intentions (or hopes), the increasingly overt theatricality of these behaviors and their shopwindow framing opened possibilities for the recognition of the mediated nature of social relations, of the entanglement of the public and the private. To one New Yorker, everyone he saw on the screen looked like actors on a Broadway stage.

By the third day, the public that appeared in the window screens of the department store, and the high-cultural center had been reduced to one of predominantly personal interaction. As discussed, friends and family members arranged to meet in front of the camera to reconnect from opposite ends of the country. This development was in many ways very telling, reflecting the privatization of public space and media on a personal level, in addition to and deeply intertwined with its commercialization and commodification. The news coverage, meanwhile, treated *Hole in Space* like a playful anomaly, a cute artistic stunt. At the end of the brief ABC News segment, the anchors back in the studio mocked the work by waving to each other from across the desk with telephones in hand (figure 2.9):

ERNIE: Bruce, I can see you, can you see me?

BRUCE: Hey, what's the weather like over there?

ERNIE: The weather isn't bad.

BRUCE: That's ridiculous, get off!

This re-mediation represented the project as a frivolous thing that artists do, instilling an aura, the exact type of categorization and recognition value that Mobile Image was attempting to disrupt with regard to notions and tools of self, site, and sight.

Deutsche describes that kind of destabilization of normative positions and perceptions in distinctly feminist terms. The introduction of conflict, division, and instability to the norms of proper social identification and behavior presses viewers to, as she puts it, "examine their own role and investments in producing images" and, by extension, their role in consuming them.[122] In her discussion of the 1982 exhibition *Public Visions*, which featured works by Sherrie Levine, Laurie Simmons, Cindy Sherman, Nancy Dwyer, and others, Deutsche presents the interrogation of unexamined modernist *positionality* of the subject vis-à-vis the object, along with the action, mechanisms, and devices of that positioning as a challenge to the early 1980s discourse of postmodernism in art and its "indifference to sexuality and gender." Such interrogations draw attention to what is invisible in the visual field—"the operations that generate the seemingly neutral

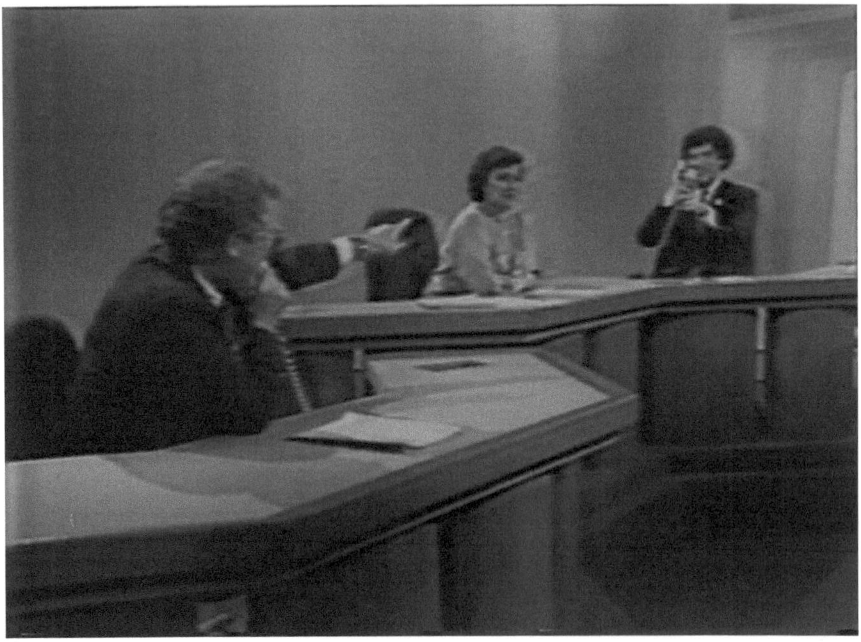

spaces of the image and the viewer."[123] Akin to the experience of *Hole in Space*, viewers of the aforementioned artists' work are called out, identifications with viewing positions spelled out, as the "image 'sees' the viewer." Sherman, for example, in her ongoing project of performing a range of character types sourced from mass media, explores roles and characters "not as reproductions of real identities but as effects produced by such visual signifiers as framing, lightning, distance, focus, and camera angle."[124] Likewise, Kruger's work, particularly her use of personal pronouns, explicitly acknowledges the viewer's presence, "erod[ing] the invisibility that protects the purportedly neutral viewing subject from interrogation."[125]

2.9

ABC News segment on *Hole in Space (A Public Communication Sculpture)*, screenshot from video documentation, 1980. Courtesy Rabinowitz/Galloway Archive.

Before the shopping windows at Lincoln Center and the Broadway department store, mediated images of others, and of self among others, similarly looked back, becoming images performed, eroding the invisibility of public and private expectations of how viewing subjects are constituted. As with Kruger's confrontational work, those depicted on the screens of *Hole in Space* repeatedly called people out, demanding that onlookers become visible, while continuously drawing attention to the commercial apparatus. *Hole in Space* juxtaposed and assembled the devices and mechanics of advertising and shopping, of cinema, urban exploration, and techno-futurist communication, without offering a preconceived resolution or premeditated expectations of where these elements belong in the public sphere or what kind of orientation they normally or ought to afford. There was very little integration of individuals into a more utopian or ideal community as the voices of women and minorities remained marginalized even when present, while others uncomfortably and self-consciously pressed themselves to the foreground. Integration and inclusion amounted to a usurpation of difference into a community of consumers, a reification without a reorganization of experience.

In Bratton's terms, the City consists of surfaces and interfaces, passages and partitions, makers and signposts, roving subjects and objects. It is made up of and traversed by lines and grids and networks that organize and channel the flow of bodies, goods, and information. The City directs as much as it narrates; it is both descriptive and generative. At its most basic, *Hole in Space* linked two outdoor locations and audiences in two different locations across the country. As such, however, it also linked the complex systems that govern those places and people and their interactions. The work connected public and private; individual bodies, representations, and social formations; mechanisms of access and exclusion; commerce and culture; ritual, play, and consumption; space, place, and site.[126] And, as mentioned, it involved divisions as much as unifications, uncritical euphoria as much as critical consciousness. The User was thus situated within and as part of a network of connections and disconnections, engagements and disengagements. *Hole in Space* attracted a relatively homogenous audience, partly due to its physical locations, which grew out of the discussed "uneven development" and gentrification of urban space in New York and Los Angeles that increasingly turned such spaces into the antiseptic social sites that found their correlate, as discussed by Everett and Nakamura, in the deracialized new media networks. But, crucially, the work addressed its participants primarily as private individuals, at once individuating and dispersing them, (re-)*interpellating* them as *Homo economicus*, as what were soon to become *Homo persona*, subject positions availed, chosen, and assigned according to profiles of consumption (whether of goods, attitudes, or identities), as what Bratton refers to as "shadows of the personified simulations of ourselves."[127] This was not a renewed coercion into an existing mold but rather the negotiated organization of potentially volatile self-positionings and -imaginaries, a citizenship that is defined by discretion and disembodiment, the mobility of which is privileged and rewarded in the "Copernican trauma" of the complexity of the world as the Stack. The city dwellers in *Hole in Space* played with the mechanisms that connect the Stack's layers and platforms, but in the end, they did not make space for anyone other than the "User-citizen-laborer-customer."[128] If, according to Dourish, embodiment is primarily relationality, a manifestation of how things are embedded in the world, then *Hole in Space*'s public assembly of private individuals formed a body politic that actively disengaged from questions of race, class, and gender as shared and contested modes of past, present, and future social formations. The prevailing logic of connectivity, by which the complexity and ambiguity of experience as relationality between and among individuals and the social is relegated to the private, begets a homogeneity through individuation.

Yet, despite this reaffirmation of discrete selfhood, what *Hole in Space* did offer, not only to its User participants but also as an artwork to be studied in relation to past and future aesthetic iterations of platform relationality, is a *staging* of the Stack. Over the period of three days, Users wrestled with the interface and with interfacing, with how to negotiate the available technology as a means to connect and disconnect in multiple ways and formats. *Hole in Space* thus engaged a range of associated concerns: presence and absence; various media and mediations as framing, screening, and mirroring; individuals and constellations of individuals, static and flowing, site-specific and dislocated; site and space as physical and immaterial nodes; people as actors, bystanders, citizens and noncitizens compliant or, on occasion, deviating from ritual and play, affirming and challenging through presence and through absence the politics of the polis (of the polis as *territory*), of how subjects and objects (and subjects *as* objects, and vice versa) are shuttled or throttled both through concrete and electromagnetic fields and through ideas, narratives, and images. Most importantly, however, as *Users*, in Bratton's as well as the Russian Constructivist theorist Boris Arvatov's sense, those directly participating in the work or employing it, in turn, as a frame and perspective, as an artistic and historical technology, arguably experienced subjecthood as relative, as "cobbled," rather than total, whether individual or social. Users saw themselves mirrored in the homogeneity of their preconceived differences, the limited ability of this reach across time and space to truly diversify the assembly, its behaviors, and its perspectives. What was made available to be sensed in this case was not the technical dimension of the apparatus or the financial means that facilitated the work but rather the technological as aesthetic, as prosthetic. The crucial intimacy of encounter resided less in the moving reunions of friends and family members across the country than in the "touchlessness" of *any* encounter, whether between people or objects, via screen or in person. *Hole in Space* engaged both the myth of technology as "antiseptic" and its refutation—the idea of technology as mediation, as a form of touching. For better or worse, the work collapsed what Bratton calls the "dichotimization of interpersonal and infrastructural modes of sensing."[129] "User sovereignty," he explains, comes from the understanding, politicization, and acting upon the realization that "mediation is not a secondary condition of our embodiment, it *is* the condition of our embodiment."[130]

Hole in Space functioned as a display of the norms and possibilities of publicity, an experiment in a technological reply to the transformation of urban public space, a connectivity that tested the limits and potential of connectivity, encounter, and positionality, to reveal the general social horizon of experience as an always-already contested construct. The romantic model of flânerie found an extension in the equally deromanticizing presentation and performance of interpersonal telecommunicative exchange as commodified ideals of bourgeois pasts. Shopping for relationships in the urban and mediatized spaces as theaters of purchases, participants and artists alike were able to observe the insufficiencies of the products on view as well as the positions they afford their customers, the screen less a mirror for affirmative ego-casting than a mediated, reified stage for the performance of social relation through a technical apparatus—in its surprise and joy, playfulness and potential, as much as its articulation of the relation between material utility and ideological projections of progress and communal formation through public interaction. Technological reproduction does not automatically avail new "healthier" structures of communication and participation in social space, but neither does it inherently subvert mass cultural forms of production and distribution. Yet if, as with *Hole in Space*, it is arranged as a tool of multiple, sometimes cohering, sometimes competing applications, as something perceived as a thing that both regulates and enables perceptions, as an instrument of dreams as well as history, then the apparatus is presented as a potentially transformable and emancipatory organizer of experience. This reorganization remained latent in *Hole in Space*, as the work did not yield Jane Jacobs's fabled "sidewalk" or Manuel Castells's place "in-between" nodes, but rather what Miwon Kwon describes as "wrong places," those "likely to expose the instability of the 'right place,' and by extension the instability of the self."[131] Yet, it previewed what a new cooperative space could look like once roles, attitudes, and subject positions are understood as and through the tools of mediation, their ideological function and potential re-functioning. As Alexander Kluge puts it, "Cooperation here implies the exact opposite of those neoliberal structures of collaboration where all individual powers are bundled and committed to a predefined objective in the name of efficiency," whether that objective is producing new markets or reinscribing social and economic parameters of ownership and control. "Instead, it involves forms of exchange in which *space is made* precisely for the unexpected, random, and resistive, in which 'ego gates' are lowered to give rise to a third facet between the negotiating parties."[132] This making of space would happen four years later, in *Electronic Café*.

3

ELECTRONIC CAFÉ: TECHNOLOGICAL COUNTERPUBLICITY

Federal elections, Olympic ceremonies, the actions of a unit of sharpshooters, a theater premiere—all count as public events. Other events of overwhelming significance, such as child-rearing, factory work, and watching television within one's own four walls, are considered private. The real social experiences of human beings, produced in everyday life and work cut across such divisions.
—Oskar Negt and Alexander Kluge, *The Public Sphere and Experience*[1]

Hosted by the city of Los Angeles, the 1984 Olympics was a peculiar mix of ritual celebrations of human achievement and harmony, spectacular displays of technological prowess, and omnipresent corporate sponsorship (a first in the history of the Games), pointedly animating the struggle that is the public sphere. The highly publicized pomp and circumstance of brotherhood and progress clashed with the socioeconomic realities and perceptions of a deeply segregated city and its constituencies. At the height of the Reagan-era's conservative backlash, neoliberalism's utopia promised greater freedom and liberty through increased privatization in terms of free markets and corporate control, as well as personalized forms of labor, leisure, and responsibility.[2] The public sphere and its apparatuses of publicity do not exclude the private but rather organize it as part of an overall experience, socially binding and economically viable. As discussed in the previous chapter, Los Angeles, the "city of the future," had long been a site of such organization, where urban development meant the bulldozing, displacement, and defunding of entire neighborhoods in the name of common progress and the violent marginalization of communities and their (self-)perception in relation to the nexus of hegemonic values, behaviors, and imaginaries. The public sphere comprises such organization and relationality, along with contradictions between ideals, expectations, and everyday life. Again, it is a site of struggle over psychological, emotional, ideological, and technological conditions as much as material conditions. Negt and Kluge's words (previously cited in this book's introduction) are worth repeating: "The public sphere denotes specific institutions and practices (e.g., public authority, the press, public opinion, the public, publicity work, streets, and public places); it is, however, also a general horizon of social experience, the summation of everything that is, in reality or allegedly, relevant for all members of society. In this sense publicity is, on the one hand, a matter for a handful of professionals (e.g., politicians, editors, officials), on the other, something that concerns everyone and realizes itself only in people's minds, a dimension of their consciousness."[3]

With *Electronic Café* (figure 3.1), Mobile Image chose to participate in this contest over what (and who) is perceived as relevant to society. Their third major project modeled a different kind of utopia, one forged precisely by engaging the contradictions manifested by the dynamics of public and private life. Modeling a reflexive and critical form of fantasy, the work culminated the group's past decade of research and experimentation, fully realizing a heterogeneous, self-organized counterpublic experience.

Electronic Café linked public locations in five very different Los Angeles neighborhoods: South Central (now called South LA), East LA, Koreatown, Venice Beach, and Downtown LA (figure 3.2). Each site in this "telecollaborative network" was equipped with an array of communication devices that were futuristic for the time—a computer messaging system, searchable text and pictorial databases, image exchange and audio-conferencing equipment, slow-scan television cameras, digital writing and drawing tablets, and high-resolution printers—all made available to local residents. Visitors could post messages and images, participate in conversations, appropriate and manipulate material from both mainstream media and users at other café sites, record and archive memories,

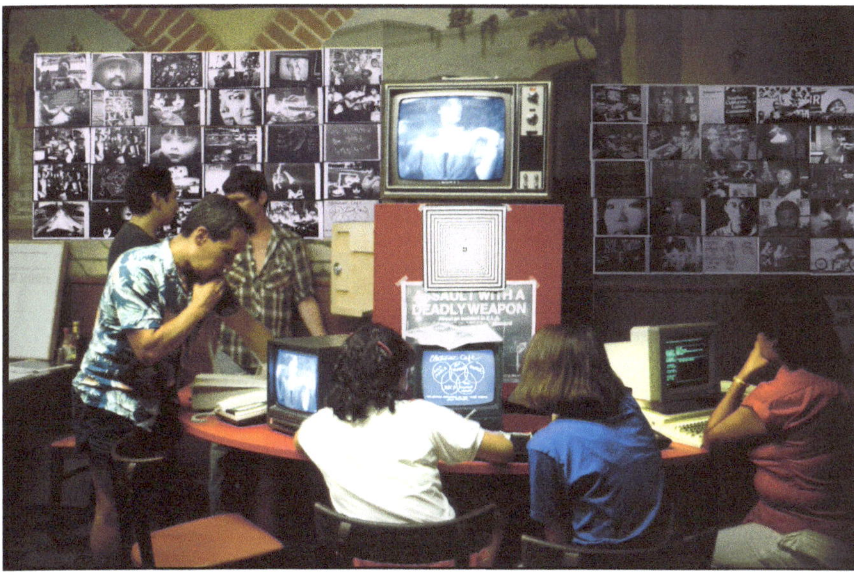

3.1

Mobile Image, *Electronic Café*, Ana Maria site (East LA), 1984. Courtesy Rabinowitz/Galloway Archive.

histories, and encounters, and search and retrieve contributions from the expanding database. Operating six hours a day for seven weeks, the project encouraged participants to use state-of-the-art telecommunication technologies to engage critically the material and immaterial connections and divisions that existed across this notoriously sprawling and fragmented city.

Connecting four geographically, culturally, and ethnically distinct neighborhoods and a public art institution through such technologies, *Electronic Café* was a fully fleshed-out model of critical and productive fantasy in Negt and Kluge's sense of the term. Rather than offer a spectacular vision of a fanciful alternative reality divorced from established

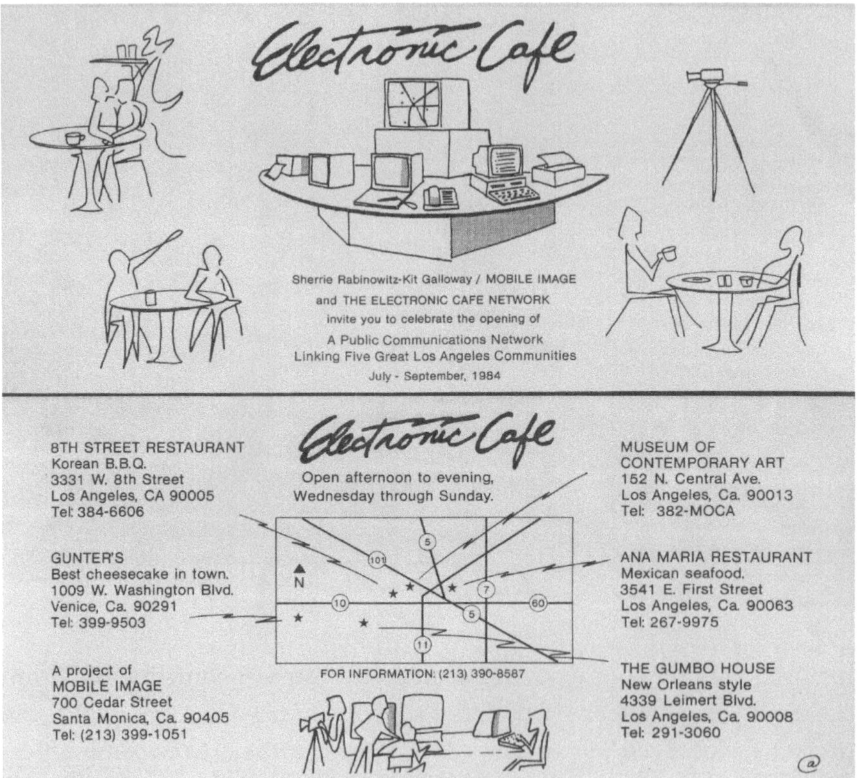

3.2

Mobile Image, announcement card for *Electronic Café*, 1984. Courtesy Rabinowitz/Galloway Archive.

norms and limits, this fantasy is built upon those very norms and limits, along with the histories, conditions, and latencies that determine and are determined by them. Inspiring a form of imaginative speculation that is confined to neither sweeping progressivist ideals nor hyper-personalized dreams, *Electronic Café* enabled the formation of temporary, overlapping, substantive relations that linked the apparent incompatibilities of social and individual experience, of public and private. The technologies used and shared did not romantically transcend the city's and its population's divisions, and yet the project also did not succumb to merely reiterating structures of dissociation and confinement. In many ways, the work took as its point of departure the constraints or limitations recognized through or consciously built into *Satellite Arts* and *Hole in Space*. In the former, through its lags and glitches, televised dancers remained visibly tethered to their physical bodies and spaces, showing that there is no autonomous "third" space, that although not confined to its corporeal limits, being and experience are always already mediated. *Satellite Arts* also made palpable the difference between broadcasting and communication at those moments when the televisual encounter was more than an individual unidirectional extension into the electronic ether. *Hole in Space*, in turn, put to the test Mobile Image's vision of a "public telecommunications sculpture," demonstrating the limits of "the public sphere" as a place for the autonomous and productive discursive encounter among informed citizens. Although situated in urban outdoor space, the coast-to-coast audiovisual interaction was weighed down by the increasing privatization of experience, both commercially and individually. The work's nuances—its pointedly chosen and juxtaposed sites of high culture (Lincoln Center) and blatant consumerism (Broadway Mall); the screens' allusions to public service announcement, entertainment, and shopping windows; the reduction of electronic encounter to the spectacular performance of stereotypical behaviors and exchange of personal messages—seemed to illustrate Negt and Kluge's claim that, on many levels, the public sphere is "the aggregate of individual spheres that are only abstractly related."[4]

Electronic Café was conceived and realized in the midst of early 1980s debates regarding material and immaterial ownership, access, and participation at a moment of deep suspicion toward all forms of collective agency. Neoliberal ideology had bastardized and institutionalized the anti-authoritarian and decentralizing impulses of the 1960s emancipation movements, while the 1970s financial crisis was presented as a failure of municipal and state fiscal discipline, ushering in wage freezes, cutbacks in social and public services, and, most crucially, an economic and ideological swing toward private funding and initiatives. As David Harvey observed, these tendencies, in the United States and elsewhere, culminated in deregulated and crude forms of wealth and power distribution, including the commodification of land and labor, the "suppression of rights to the common," and (neo)colonial and imperialist "processes of appropriation of assets," under the guise of "liberty" and "freedom."[5] The questions of who owns the tools of creation and destruction and to what ends and in whose interest are they wielded were of utmost concern. Under these circumstances, technological innovation and increased personalized access to the media, whether through a steadily growing plethora of cable television channels or the introduction of the personal computer, were slated toward personal consumption in niche markets rather than social labor, and participation in society was largely reduced to the competitive private ownership of commodities, be they things, ideas, or identities.

Yet, the postmodern condition also begat, or demanded, the productive reconsideration of histories and perspectives that would challenge the logic of late capitalism as the necessary outcome of the turmoil caused by domestic and international liberation and social movements, along with the hegemony of modernist notions of growth, productivity, and selfhood. Feminist and critical race theorists such as Donna Haraway and bell hooks advocated for strategic subject positionings outside of binary identity structures, while philosophers such as Nancy Fraser and Michael Warner were "rethinking the public sphere," "the mass public and the mass subject," and Negt and Kluge's seminal ideas concerning counterpublic resistance and the persistence of class-based exploitation were translated, circulated, and discussed in prominent art and academic journals.[6] Urbanists, including Raul Villa, Rosalyn Deutsche, and Edward Soja, traced the fight over urban imaginaries in US metropolises through mapping, zoning, and spatial design as devices of social control; Manuel Castells and Saskia Sassen pondered the overlapping and interdependent spheres of technological-information networks and the geographic-material flow of bodies, goods, and services.[7] Along with Suzanne Lacy,

ASCO, and others discussed previously, artists such as Ulysses Jenkins engaged LA-specific struggles over psycho-geographic territory and the media as an apparatus of framing and managing discursive and performative space. In 1982, following the efforts of feminist artists in (then) new media, Sylvia Harvey advocated for a reconsideration of avant-garde models of artistic production rooted in the belief in socialism as a viable political option, in "scientific thinking" as a critical-artistic practice, and in the validity of creating a Leftist popular culture, notions distrusted or outright rejected in canonical Cold War art histories.[8]

As they progressed in their work, Galloway and Rabinowitz's plans for *Electronic Café* were increasingly driven by similar questions of social transformation, the politicization of technology as apparatuses of production and distribution, and the potential of redefining critical collective engagement. They were concerned with ownership of ideas and information, access to devices and networks, and the usefulness of their project as an experiment or model of social and ideological innovation. The artists were keenly aware of the difference between symbolic (and commercial) novelty and substantive change in relationality—the way in which individual and social subjects frame and organize their connections and experiences with and of one another and their environment through technology/tools. In their archive, a folder titled "EC Original Development Materials" includes notes collected during planning sessions and discussions with peers and collaborators in 1983. In one section, Galloway and Rabinowitz acknowledge: "What is feasible technically, however, is a very different question from what is socially and institutionally feasible. For example, much of the application of information tech results in decentralization and often a flattening of the hierarchy of power. Desirable though this may seem in theory, in practice it ignores 2 realities. Many people, through instinct, education or attitude do not relish responsibility, while others may prefer to retain their positions of power in their status quo."[9] A list they compiled under the title "Philosophy/Value System" includes the following:

> What constitutes the public interest?
> Distinguishing between public and private information.
> Distinguishing between free and restricted information.
> How will success or failure of prototype affect general public attitude?

> Who is the audience, what is the public dividend, what is the product being sold and to whom?
> Can this demonstrate that human effectiveness can be enhanced via man–computer communication?
> The democratic process?
> Will the prototype accelerate the extension of experimental methods
> Transform individual intelligence and social effectiveness?
> The real issue is that of allowing access to the whole picture of what is going on.
> It takes more than mere electronics to make a network function.

And some of the "issues" they explicitly planned to address with *Electronic Café* include:

> Information rich—information poor
> Autonomous realities
> Communities of Consciousness
> human scale—technological scale
> privacy and electronic invasion
> art in a technological society
> healthy technological models and prototype
> cross-cultural communication
> appropriate technologies
> survival skill in an information-intensive environment
> aesthetics research[10]

Although Galloway and Rabinowitz did not couch their project in the traditional language of class antagonism, their philosophy, inquiry, and practice recalls Negt and Kluge's examination of the history of the public sphere through the lens of ownership, production, and distribution. As Fredric Jameson points out, Negt and Kluge use the term "proletarian" in its most general sense (and as an attribute rather than the designation of a social entity): "Proletarian, i.e., separated from the means of production, designates not merely the labor characteristics of the industrial proletariat, but all similarly restricted productive capacities."[11] The latter include the making of ideas, narratives, and identities, and, crucially, the transformation of potential futures through latent past and present experiences. Labor includes material and immaterial production and reproduction, development as well as maintenance work in all spheres: public, private, and otherwise, begetting questions of value, interest, and usefulness of such work for particular constituencies and publics in relation to the general horizon and to one another. Negt and Kluge emphasize the political need to employ what seemed to be outmoded terms:

> The word *proletarian* has . . . taken on an attenuated, indeed an anachronistic sense. Yet the real conditions it denotes belong to the present, and there is no other word for them. We believe it is wrong to allow words to become obsolete before there is a change in the objects they denote. . . . It is not our intention as individuals to replace historically evolved key concepts that denote unsublated real circumstances and do not have a purely definitional character. The fashioning of new concepts is a matter of collective effort. If historical situations really change, then new words come about too.[12]

Sanctioned by modernist myths of progress and humanism, innovations in technology and consumption often (poorly) mask, and even further entrench, the ongoing divisions between those who have the power to wield the tools of perception and subjectivity effectively in a regulatory and profitable manner and those who do not. And although, as will be discussed, *Electronic Café* was conceived to entice its publics' productive imaginations with a then-futuristic-looking design, the artists were aware of the fact that what needed to change were the processes of production and utility rather than just the devices employed. As they put it, "In the end the technology used to supply the means of communication doesn't really matter."[13] Neither art nor technology should be functioning as symbolic liberation from ideology, as commodities making up for a lack of or blocking, in Negt and Kluge's words, "any genuine coherence" between what is otherwise conveniently differentiated between public and private, between what is designated as relevant to (and thus to be complied by and aspired to) society as a totality and what is relegated as being of merely private, meaning personal interest, the latter affecting any and all of those experiences that cannot be subsumed under the prevalent images and performances of normative bodies, needs, and actions.[14]

Recalling the traditional avant-garde's rallying cry of reconnecting art and life, image and experience, Mobile Image essentially provided a technics of aesthetics akin to what Boris Arvatov had termed "socialist objects." Writing in the early 1920s, Arvatov expressed his admiration for the technical development and industrial production under capitalism, but he was also highly critical of the kind of things made available to its masses. In order to achieve a truly popular culture (one not made *for* but *by* the people) that would unite the ideal and the everyday, the public and the private, both the means of production and the objects produced would have to cease to be commodities, hence, things containing symbolically what was actually lacking, gratifying only superficially real existing needs. Arvatov advocated a new "culture of things" to overcome the dualism of *bytie* ("associated with the spiritual, the literary, and the transcendent") and *byt* ("the mute, material, and tradition-bound"), a binary dynamic that was a historical artifact of class division, wherein the concept of consumption was created in opposition to that of labor.[15] A proletarian culture would break down this dualism through the "active agency of socialist things": rather than commodities—passive objects that serve as *substitutes* for relations between producers, things, and world—socialist objects would be tools animated by their use value for social labor (rather than by exchange value and private-property relations) responding "to the social needs of the everyday life of [the] historical moment."[16] Crucially, given Mobile Image's concern with both communication and its products/outcome (along the aforementioned lines of Negt and Kluge's definition of proletarian), the socialist objects at stake in 1984 included not only the technical equipment made available but also ideas, norms, dreams, histories, and subjectivities *as technologies*—that is, as devices constructed and wielded in relation to and by power, and subject to struggles over ownership. A futuristic, or rather "fantastic," use of telecommunication technologies would have to yield more than a reproduction of existing subject positions and (material and immaterial) property relations. Simply connecting urban neighborhoods and communities with technical gadgets would not suffice. As Galloway and Rabinowitz's put it in their 1983 Olympic Arts proposal for *Electronic Café*: "Though the real relevance of satellite communication is crossing oceans—connecting cultures and nations—serious questions are now being asked about the notions of 'free flow information.' For as many developing nations rightly perceive, the flip side of the 'free-flow' is cultural imperialism. Will extending the reach of communication technology really promote greater understanding and communication, or will it only accelerate the rate of cultural homogenization already attributed to television? Up to now, the free flow of information has been a one way street."[17]

A counterpublic or proletarian technology would have to offer a new organization of experience. It would have to model the making of a new contingent yet historically rooted social horizon, a new positioning of subjects, perspectives, and knowledge, utilized not as a substitute for an existing aesthetics and technics, politics and economics of experience, but built on the expansion of the given in order to transform its very logic, of the relationality of the public and the private, the total and the partial. According to Haraway, such a technology would have to grapple with the problem of how "to have *simultaneously* an account of radical historical contingency for all knowledge claims and knowing subjects, a critical practice for recognizing our own 'semiotic technologies' for making meanings, *and* a no-nonsense commitment to faithful accounts of a 'real' world."[18] As the production, dissemination, and utility of information and desires, telecommunications will not "solve" the seemingly irresolvable conundrum of the general horizon (or institutionalized knowledge and apparatuses as prescribed objectivity) and the multiplicity of the local and of critically postmodernist difference. Yet, Haraway argues, "we do need an earthwide network of connections, including the ability to translate knowledges among very different—and power-differentiated—communities." We need to know "how meanings and bodies get made . . . in order to build meanings and bodies that have a chance for life."[19] If the public sphere is traditionally discursive and everyday life is performative, then counterpublicity is both: language and embodiment.

Electronic Café provided its users with narratives and images, verbal and written conversation as well as visual and sensory expression and exchange—keyboards, writing pads, and databases; mass media pictures, recorded performances, and drawings. To Haraway, vision is key because of its embodiment, and especially when the "instruments of visualization" can connect and organize "situated knowledges" outside the binary logic of relativism and totalization. No knowledge or horizon is total, and the disembodied eye of modern culture, with its comfortable and controlling view from above, is a powerful technology of prescriptive public sight. Like Haraway, Mobile Image looked not for an accumulation of "more real" or "truer" experiences on the ground, a privileging of the "vantage points of the subjugated"; there are no "innocent" perspectives, only already mediated ones.[20] Proletarian publicity means to learn *how* to see from embodied positions, from those partial and emphatically located perspectives that use technology for the fantastic exchange and organization of new and evolving collective experiences. The politics and technics of seeing in *Electronic Café*—of sourcing, manipulating, and creating images; of sending back and forth notes, poems, and anecdotes; of collecting and constructing perspectives, histories, and artifacts, stored and accessed under old and new categories in an electronic, accessible archive—presaged Haraway's "doctrine and practice of objectivity that privileges contestation, deconstruction, passionate construction, webbed connections, and hope for transformation of systems of knowledge and ways of seeing." Public and private, the generally prescriptive and the regulatory in particular, are intimately interwoven: "All these pictures of the world should not be allegories of infinite mobility and interchangeability but of elaborate specificity and difference and the loving care people might take to learn how to see faithfully from another's point of view, even when the other is our own machine."[21] Learning how to see, to understand and wield one's own standpoint as a technology of positioning and being positioned, and to experience and strategically employ the continuous partiality of subjecthood in order to create new vantage points, chart new territories, and build new solidarities, these were the goals of *Electronic Café*. Fantasy—again, in Negt and Kluge's sense of the term, as the productive alignment of specific social needs and desires, critical consciousness, and the outside world—was the organizing force of a technological counterpublic experience in which the city of Los Angeles became a site of struggle over territory, imagination, self, and community.

ELECTRONIC CAFÉ

Electronic Café was part of the Olympic Arts Festival, a citywide celebration of "international brotherhood" meant to exemplify the spirit of the Olympic Games.[22] Sponsored by the Times Mirror Company, the ten-week event included more than four hundred works by artists and performance companies from every continent and was emblemized by Robert Graham's monumental neoclassical *Olympic Gateway*, a post-and-lintel structure topped by two bronze nudes, commissioned for the entrance to the Los Angeles Memorial Coliseum. According to Director (and then Cal Arts President) Robert Fitzpatrick, the festival was conceived "with a Greek verb and a promise. The verb is *thaumadzo*—to be seized with wonder, to experience awe, to be surprised and to take delight in discovery." As Fitzpatrick explains, the festival was based on the premise that "art is not a form of propaganda but an instrument of truth, an opportunity to put aside differences and rejoice in being alive," and participating countries explicitly agreed not to "preach."[23] Promotional materials gathered by Galloway and Rabinowitz in advance of their project reiterated the message of art as an agent of global harmony, declaring that "the arts belong to everyone" and that "in an increasingly multicultural society, the arts provide a universal language, a bridge of illumination that connects all of us."[24]

Electronic Café was unusual in its confrontation with the relational politics of identity, place, and space in present-day Los Angeles. Although its focus on exciting new forms of cross-cultural exchange may have seemed compatible with the goals of the Olympic Arts Festival, as well as the games themselves, it distinguished itself amid a citywide spectacle that promoted supposedly universal classical principles but generally avoided—and arguably distracted from—the real conditions of the city as a site of economic, social, and political struggle. In a letter supporting Mobile Image's proposal to the festival, Samuel Mark, Acting Director of USC Institute for Hispanic Media and Culture, recognized this distinction: "The 'Electronic Café' has my enthusiastic support for two principal reasons. One, that it will reach some of the many different ethnic communities that compose our City, and two, that it utilizes computer-video installations in an innovative and artistic way. With the exception of the 'Electronic Café' and very few other presentations, I am extremely disappointed with the 1984 Olympic Arts Festival because its organizers have not done enough to reach and represent the Hispanic community and other communities, nor to present innovative cultural manifestations.[25]

Mobile Image designed *Electronic Café* simultaneously to inspire future imagining and to place such imagining in dialogue with the

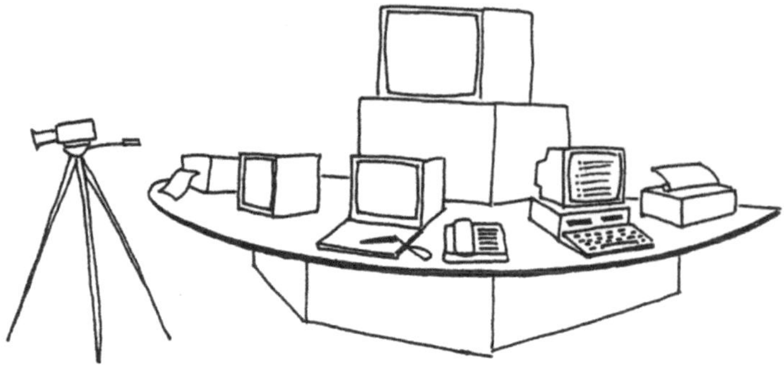

present. At each locational node, the available telecommunication equipment was set within a self-contained semicircular console (figure 3.3), whose streamlined, monochrome design and precise arrangement of devices—several small monitors and workstations below, with a large screen overhead—evoked control panels featured in countless science-fiction movies and television shows. Yet, Mobile Image took care to prevent its sci-fi aesthetics and machinery from operating in isolation: the network was emphatically grounded in the specificity of the sites that comprised it—the various neighborhoods of Los Angeles, the identity labels conventionally affixed to them, and the particular establishments in which the futuristic consoles were installed. Four of the five locations were restaurants (figure 3.4), selected by local residents collaborating with the artists because those establishments were firmly rooted in their respective communities. The East LA node was installed in Ana Maria, an "all-Mexican" family restaurant, with waitresses wearing traditional dresses and with murals featuring colonial architecture.[26] South LA's was located in the Gumbo House (figure 3.5), a Cajun restaurant catering mainly to the local Black population. Koreatown's was in the 8th Street Restaurant,

3.3

Mobile Image, diagram of console for *Electronic Café*, c.1984. Courtesy Rabinowitz/Galloway Archive.

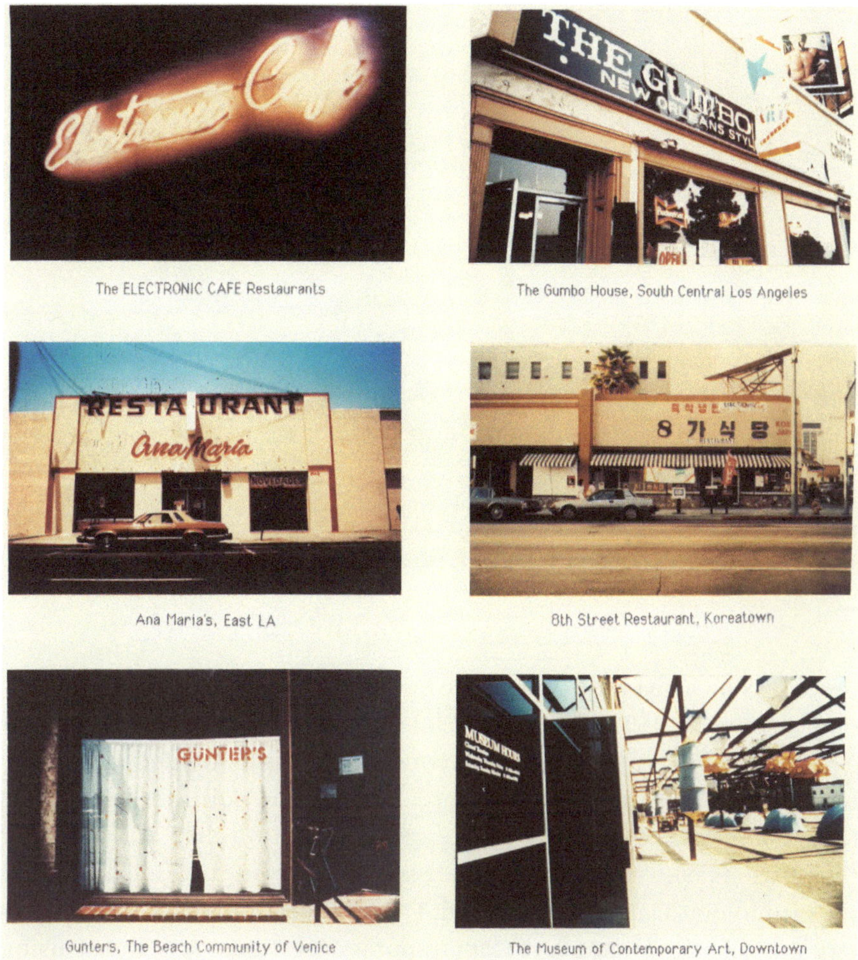

3.4

Mobile Image, *Electronic Café*, photo sheet with sign and five locations, Los Angeles, 1984. Courtesy Rabinowitz/Galloway Archive.

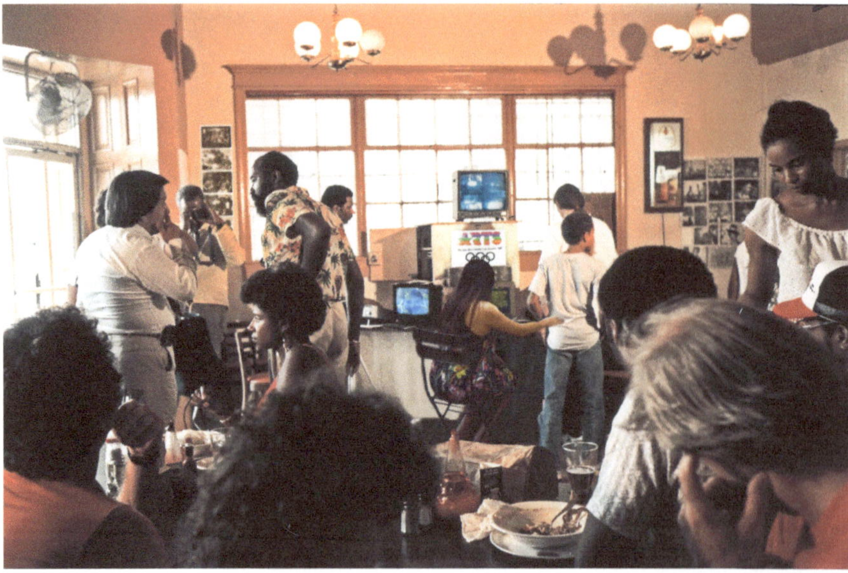

described by artist-participant Hye-Sook Park as "a very rural, Korean-only restaurant."[27] The Venice Beach location was Günter's, a bohemian eatery modeled on the traditional coffeehouse and intended, according to its owner, as "a forum for the arts and serious political discussion" (figure 3.6).[28] *Electronic Café* encouraged users to observe and consider the relations between the immediate social reality of their locations and the technological apparatus at hand. The fifth location, set in an open gallery space in the Museum of Contemporary Art (figure 3.7) and housing the network's central database, linked such realities to the very cultural institutions that maintain and are maintained by them.

Mobile Image also consciously limited themselves to then-current technological capabilities. The fantastic devices that comprised *Electronic*

3.5

Mobile Image, *Electronic Café*, Gumbo House site (South LA), 1984. Courtesy Rabinowitz/Galloway Archive.

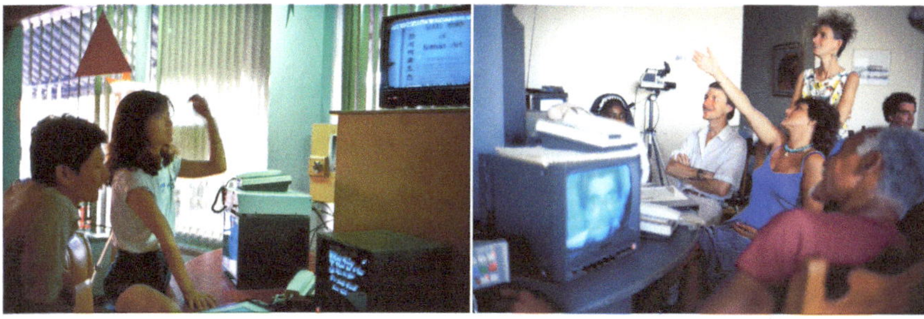

Café were state of the art but not speculative; despite their technical know-how, the artists chose to employ off-the-shelf equipment and "narrowband" networking via preexisting telephone lines rather than build new prototypes that would not be available for years (figure 3.8). Nonetheless, most of the featured equipment was still relatively unfamiliar to the general public. (*Electronic Café* excluded fax machines because they were already integrated into corporate America and thus came with preset procedures and connotations.) The point was to allow users to participate actively in the development of the still-unformed protocols and parameters of emergent technologies without getting overly distracted by or enamored with the machinery itself. Mobile Image also carefully constructed the consoles so that the complexity of the devices—the elaborate wiring and networking mechanisms—would be largely invisible, rendering the system as user-friendly as possible (figure 3.9). As Rabinowitz explained, visitors "were confronted with about $70,000 worth of equipment in each café, but the technology was transparent enough that they came away with the quality of the human experience they had."[29] Each location also had an artist-in-residence and engineer-in-residence with existing

3.6

Mobile Image, *Electronic Café*. Left: 8th Street Restaurant site (Korea Town); right: Günter's site (Venice Beach), 1984. Courtesy Rabinowitz/Galloway Archive.

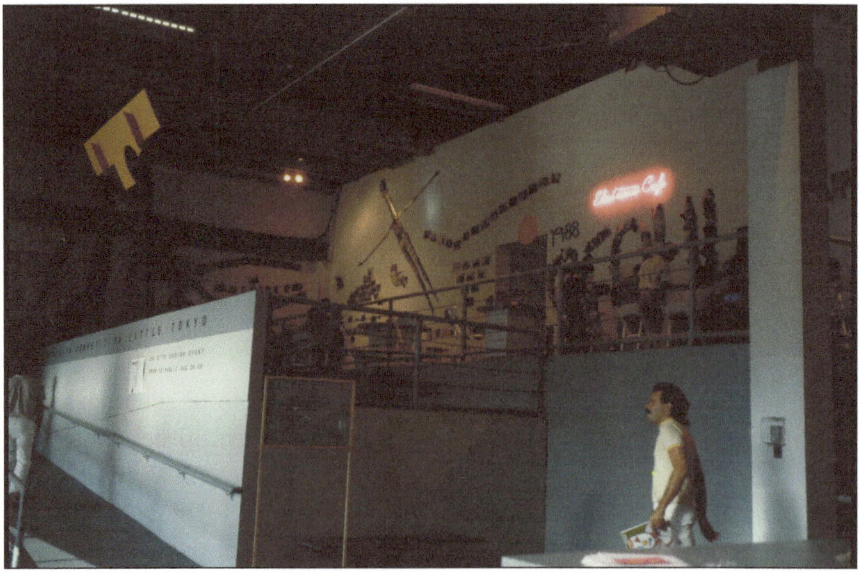

ties to the local community and the technical know-how to demonstrate the capabilities of the network, encourage participants to improvise, and help when things did not function as planned (figure 3.10). The complexity of the system—composed of high-tech devices not designed to work together—meant that lag times, periodic breakdowns, and interruptions were inevitable, and the management of such malfunctions was considered part of the work and its collaborative research model.

Undergirding *Electronic Café* was a customized network built in collaboration with the Community Memory Project, a nonprofit collective of Berkeley-based computer scientists and engineers that, over the previous decade, had been developing what they called a "decentralized, community-controlled communications system." This system would

3.7

Mobile Image, *Electronic Café*, Museum of Contemporary Art site, 1984. Courtesy Rabinowitz/Galloway Archive.

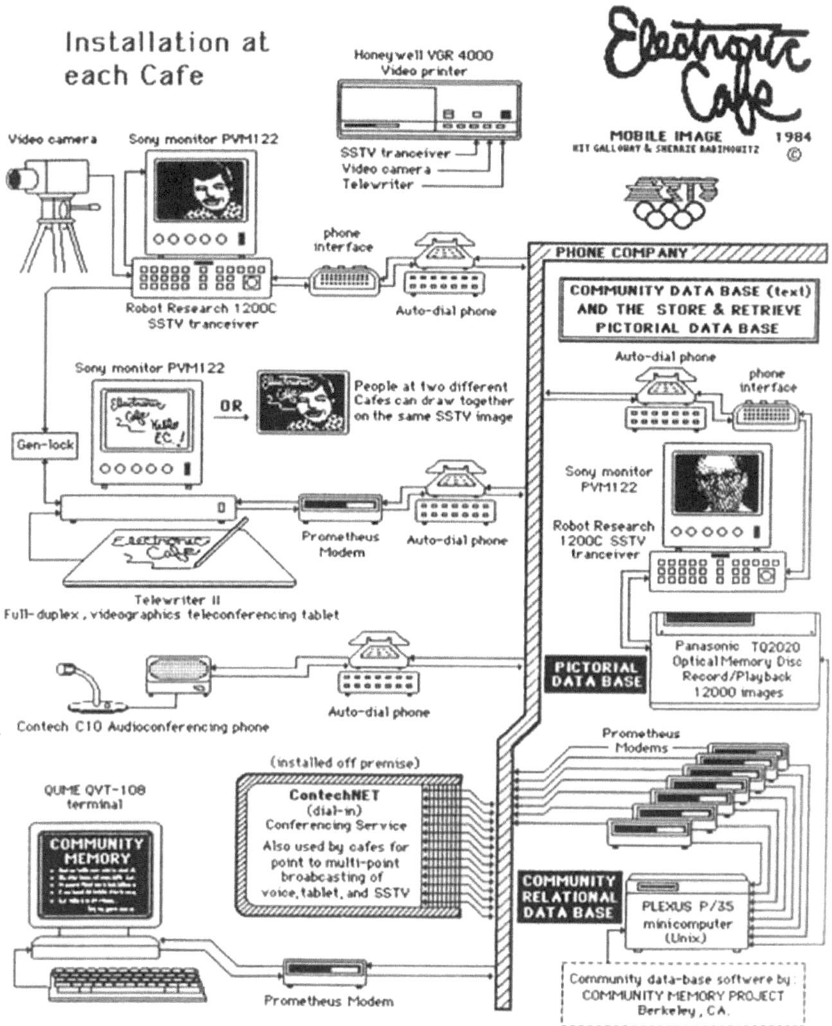

3.8

Mobile Image, *Electronic Café* setup diagram, 1984. Courtesy Rabinowitz/Galloway Archive.

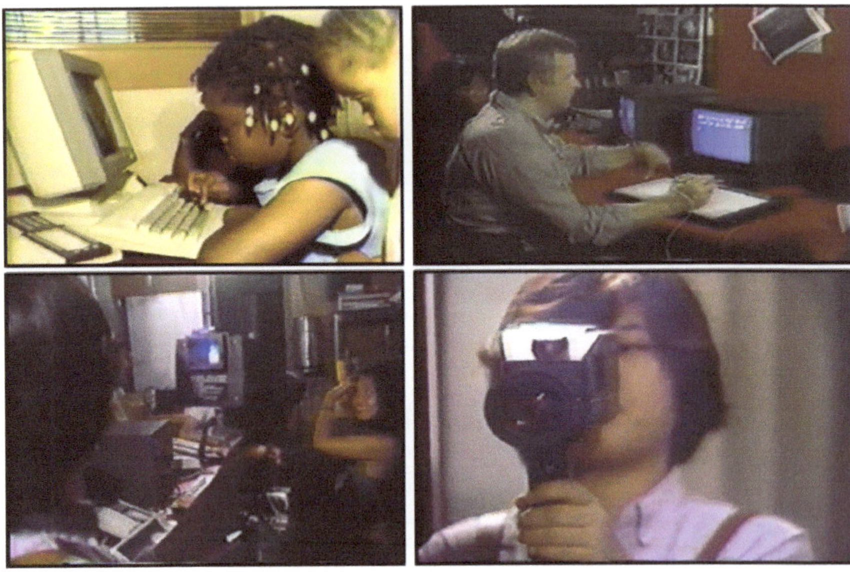

serve as "an alternative to the highly centralized information delivery by the mass communications media, which tell everybody what a few people have to say, and don't give you a chance to talk back, to talk to each other." Merging computer technologies with the politics of public access and engagement, they proclaimed that such a system could facilitate "a future in our own terms."[30] In 1973, the collective established the first public bulletin board system (BBS), consisting of three terminals at different locations—a record shop, a hardware store, and a branch of the San Francisco Public Library—connected via modem to a central mainframe computer (figure 3.11). Each terminal functioned as its own independent, interactive database, enabling users to input and retrieve information and messages according to their own needs. Operational for fourteen months, the BBSs were used extensively by local residents—for everything from carpools to restaurant recommendations to political dialogues—many of

3.9

Mobile Image, *Electronic Café*, various locations, 1984. Courtesy Rabinowitz/Galloway Archive.

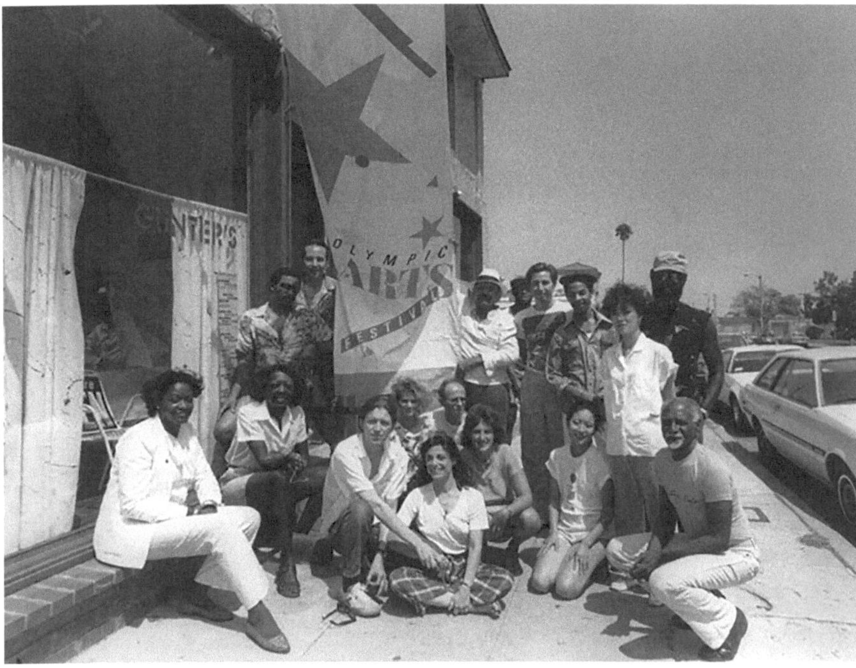

whom had never operated a computer before. As its developers noted, the experiment "showed that the public at large, without prior training, can use an electronic information exchange system to define and meet their own information needs." At the same time, those developers recognized that a network of dispersed public "nodes"—although unfeasible at the time—would greatly expand the project's potential.[31] In 1982 and 1983, they fully theorized such a network; by 1984, they were ready to construct it (figure 3.12).

Galloway and Rabinowitz closely followed this development, and in preparation for *Electronic Café*, they compiled and annotated a host of Community Memory documents as source material. Their archived notes

3.10

The *Electronic Café* team in front of Günter's, 1984. Courtesy Rabinowitz/Galloway Archive.

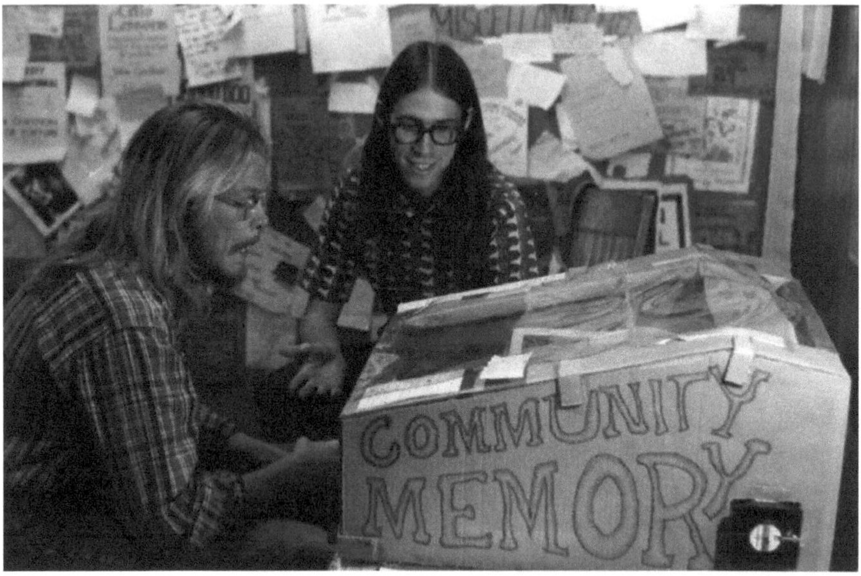

show that they were as interested in the techno-politics of Community Memory—its potential to reorganize relationships between individuals and community, private and public—as they were in its technical innovations. Under the heading "ADVANTAGES," Galloway hand copied select passages (marked here in bold) from a 1982 booklet called *The Community Memory Project: An Introduction* (figure 3.13):

Community Memory is a system for the **public management of public information**. It is an open channel for community communications and information exchange, and a way for people with common interests to find each other. It is a **shared community filing cabinet**. It is a tool **for collective thinking, planning, organizing, fantasizing, and decision making**.

3.11

Community Memory terminal at Leopold's Records, Berkeley, California, c.1975.

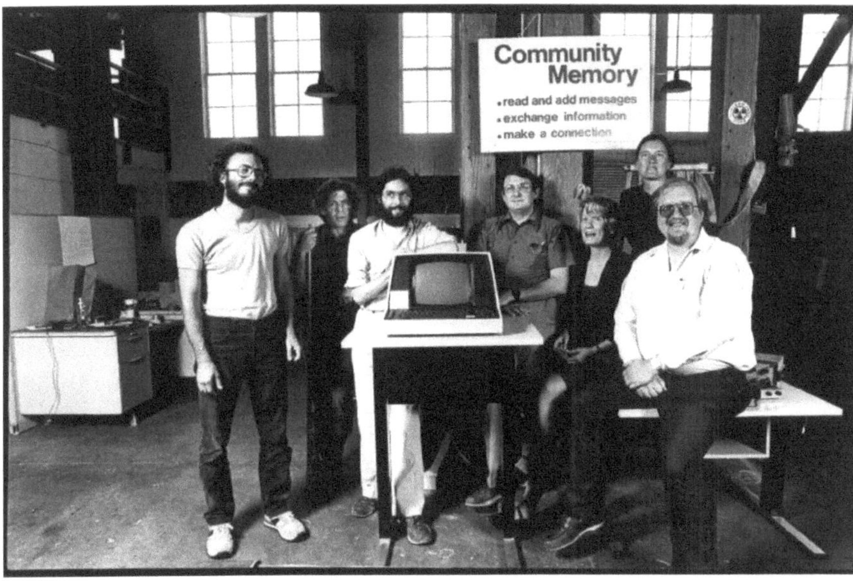

Community Memory seeks to present **an alternative to broadcast media** such as TV. The nightly national TV news—both commentary and commercials—gives people the "word" from on high, telling us "that's the way it is." Community Memory is different. It makes room for **the exchange of people-to-people information**, recognizing and legitimizing the ability of **people to decide for themselves what information they want.**

A community Memory node might also be shared by people who are working on some common project in different parts of the country—the **"community" here would not mean one geographic locality, but would represent a community of common interests.**

3.12

Community Memory group. Left to right: Carl Farrington, Michael Rossman, Phil Kohn, Lee Felsenstine, Karen Paulsell, unidentified, Ken Constad, 1984.

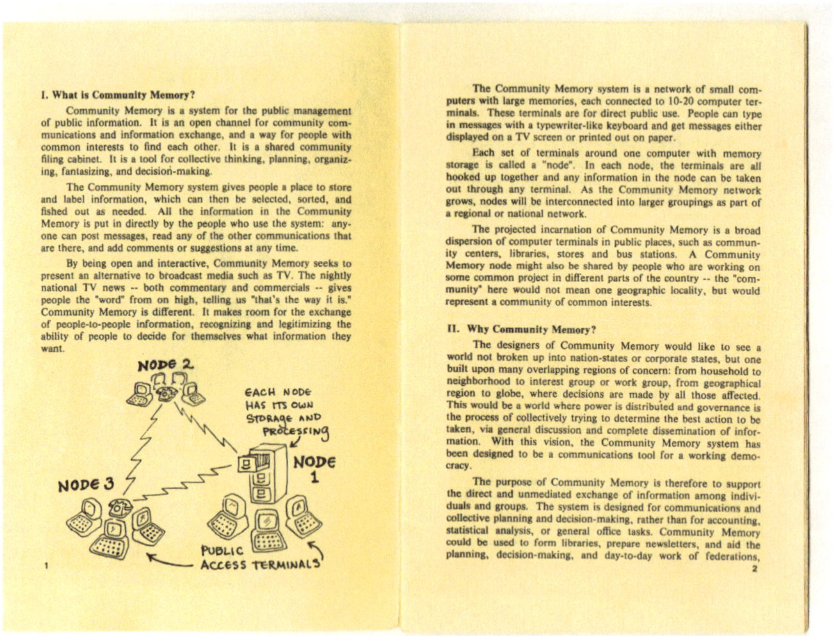

I. What is Community Memory?

Community Memory is a system for the public management of public information. It is an open channel for community communications and information exchange, and a way for people with common interests to find each other. It is a shared community filing cabinet. It is a tool for collective thinking, planning, organizing, fantasizing, and decision-making.

The Community Memory system gives people a place to store and label information, which can then be selected, sorted, and fished out as needed. All the information in the Community Memory is put in directly by the people who use the system: anyone can post messages, read any of the other communications that are there, and add comments or suggestions at any time.

By being open and interactive, Community Memory seeks to present an alternative to broadcast media such as TV. The nightly national TV news -- both commentary and commercials -- gives people the "word" from on high, telling us "that's the way it is." Community Memory is different. It makes room for the exchange of people-to-people information, recognizing and legitimizing the ability of people to decide for themselves what information they want.

1

The Community Memory system is a network of small computers with large memories, each connected to 10-20 computer terminals. These terminals are for direct public use. People can type in messages with a typewriter-like keyboard and get messages either displayed on a TV screen or printed out on paper.

Each set of terminals around one computer with memory storage is called a "node". In each node, the terminals are all hooked up together and any information in the node can be taken out through any terminal. As the Community Memory network grows, nodes will be interconnected into larger groupings as part of a regional or national network.

The projected incarnation of Community Memory is a broad dispersion of computer terminals in public places, such as community centers, libraries, stores and bus stations. A Community Memory node might also be shared by people who are working on some common project in different parts of the country -- the "community" here would not mean one geographic locality, but would represent a community of common interests.

II. Why Community Memory?

The designers of Community Memory would like to see a world not broken up into nation-states or corporate states, but one built upon many overlapping regions of concern: from household to neighborhood to interest group or work group, from geographical region to globe, where decisions are made by all those affected. This would be a world where power is distributed and governance is the process of collectively trying to determine the best action to be taken, via general discussion and complete dissemination of information. With this vision, the Community Memory system has been designed to be a communications tool for a working democracy.

The purpose of Community Memory is therefore to support the direct and unmediated exchange of information among individuals and groups. The system is designed for communications and collective planning and decision-making, rather than for accounting, statistical analysis, or general office tasks. Community Memory could be used to form libraries, prepare newsletters, and aid the planning, decision-making, and day-to-day work of federations,

2

The designers of Community Memory would like to see a world not broken up into nation-states or corporate states, **but one built upon many overlapping regions of concern**: from household to neighborhood to interest group or work group, from geographical region to globe, where decisions are made by all those affected.

As **an accessible, non-hierarchical and interactive public medium**, Community Memory will be **unique among current communications systems**. It has certain **similarities to pay telephones, public libraries, radio talk shows, and bulletin boards**, but it has the potential for being **a far more powerful tool than any of these**.

3.13

The Community Memory Project: An Introduction (from Rabinowitz/Galloway Archive), 1982.

A community can be a geographic, economic, cultural, political or recreational entity. By helping its users connect to others who share their interests and concerns, Community Memory can strengthen people's involvements in **many overlapping communities and help them work together toward common goals.**[32]

Early in 1984, Mobile Image began collaborating with the Community Memory engineers, who shared their text-based framework and helped incorporate an optical-disk database into it (figure 3.14).[33] The *Electronic Café* platform thus enabled users to post both text and pictures that could subsequently be retrieved and commented upon by other users (figure 3.15). The result was a cumulative, searchable archive that could serve as a space for public interaction and identity formation according to the particular needs of the user—not as an autonomous individual but rather as a social being with shareable concerns. It functioned as a site for collecting alternate histories and inaugurating fresh forms of political organization, enabled by Community Memory's customizable, thematically based filing scheme that could accommodate a potentially limitless multiplicity and combination of ideas and issues—high and low, public and private. The system came with a set of preprogrammed general "index words" (e.g., "music," "food," "sex") and additional words geared toward historically specific local and trans-local concerns (e.g., "housing," "nuclear," "women"), but it was also enriched by any number of user-generated categories entered onsite. Examples of these categories included "media for peace," "American Regime," "the life of a refugee," "transsexual rights," and "what it takes to be in the public." In this way, the network enabled new social formations via common and negotiated values and interests beyond preconstructed labels. It presented opportunities to produce heterogeneous relations and to recognize the power of doing so.

While demonstrating the sociopolitical potential of such relations, Mobile Image was concerned that new communication technologies were increasingly assumed to be inherently emancipatory, that the public was succumbing to the image of a spectacular future enabled by frictionless flows of information, decentralized authority, and de-territorialized relations. What was once tied to the military-industrial complex was now linked to utopian dreams of freedom achievable because of a "communications revolution."[34] As Fred Turner has chronicled, by the early 1970s,

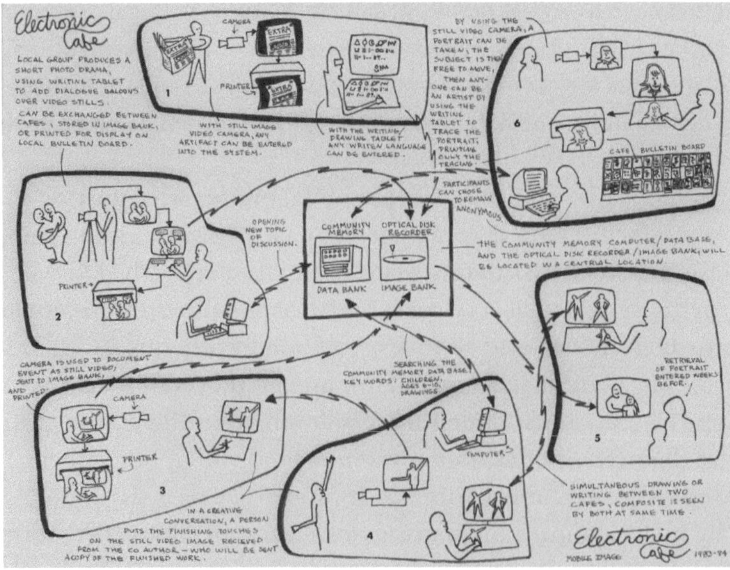

technology was increasingly seen as a social and emotional remedy: if industry and government made people "psychologically fragmented specialists, the technology-induced experience of togetherness would allow them to become both self-sufficient and whole once again."[35] By the mid-1980s, this mindset was pervasive. As Gene Youngblood noted at the time, new technologies promised to "invert the structure and function of mass media (a) from centralized output to decentralized input, (b) from hierarchy to heterarchy, (c) from mass audience to special audience, (d) from communication to conversation, (e) from commerce to community, (f) from nationstate to global village."[36] Mobile Image was highly skeptical of this promise, even as they advanced the possibility for social transformation. This ambivalence was shared by their Community Memory

3.14

Mobile Image, *Electronic Café* network diagram, with Community Memory and optical disk recorder in the center, 1984. Courtesy Rabinowitz/Galloway Archive.

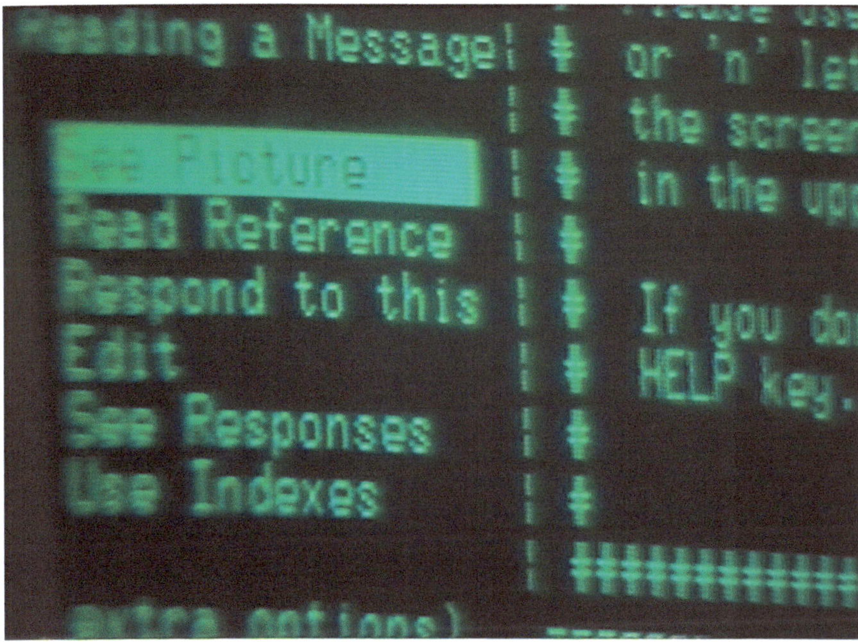

collaborators, who were conscious of the fact that, in the decade since their earliest experiments, public attitudes toward such technologies had generally morphed from suspicion and hesitation to uncritical embrace. In a 1983 newsletter, the Community Memory Project explicitly warned of the dangers of growing "'computerphilia': the notion that pushing a few keys on a terminal will liberate people, make their jobs more interesting, expand their information horizons, and give them unlimited powers"[37] The group's 1982 booklet likewise includes a section on "The Limitations of Community Memory," which warned that technology alone cannot flatten social and political hierarchies. Again, Galloway echoed these concerns by copying key passages into his own notes (in bold):

3.15

Mobile Image, *Electronic Café*, computer screen showing database operations, 1984. Courtesy Rabinowitz/Galloway Archive.

Community Memory is a demonstration of the potential of technology to be used for human liberation. However, **the existence of a potential does not assure or even make likely its utilization. Like solar energy, radio, etc., the realization of the possibilities must be accomplished in opposition to the current organization of power in the society. The structure of society will not be changed to the disadvantage of those currently holding power by the introduction of *any* new technology or application of technology . . . New technology for managing information and decision-making is a *requirement* for a *humane world*, but it can't replace the political process. Only humans can build a humane world.**[38]

What was so innovative about *Electronic Café*—and what distinguished it from the universalist doctrine of the Olympics and its arts festival as well as Community Memory's own experiments—was that it explicitly infused its futuristic demonstration of new tools of decentralization, dematerialization, and deterritorialization with the politics of identity, access, and control, place, and space. The locations linked by the network underscored the specific ethnic, racial, and class content mapped onto the highly territorial geography of the largely segregated city. As will be discussed, the selected neighborhoods were not only physically and culturally separate but also subject to much larger citywide, national, and global economic forces that at times stoked tensions between them—tensions that were aired and debated on the network. Galloway acknowledged such tensions, noting, for example, that the Black and Korean-American communities were "at each other's throats" over perceptions of economic exploitation.[39] This type of site specificity tied the technological apparatus to the complex histories of the places involved and to their relationships to each other, to power, and to a range of competing past, present, and future imaginaries.

Similar to *Hole in Space*, initial interactions within the network tended to be characterized by preexisting notions of cultural difference and expectations of how one would be perceived at the other locations. As Rabinowitz recalled: "The first broadcasts were the communities identifying themselves. Each café found it important to define their cultural identity through some kind of presentation or performance. They began to transmit images and icons and ideas that demonstrated the breadth and scope of their culture. They were very conscious of saying 'This is who we are.'"[40]

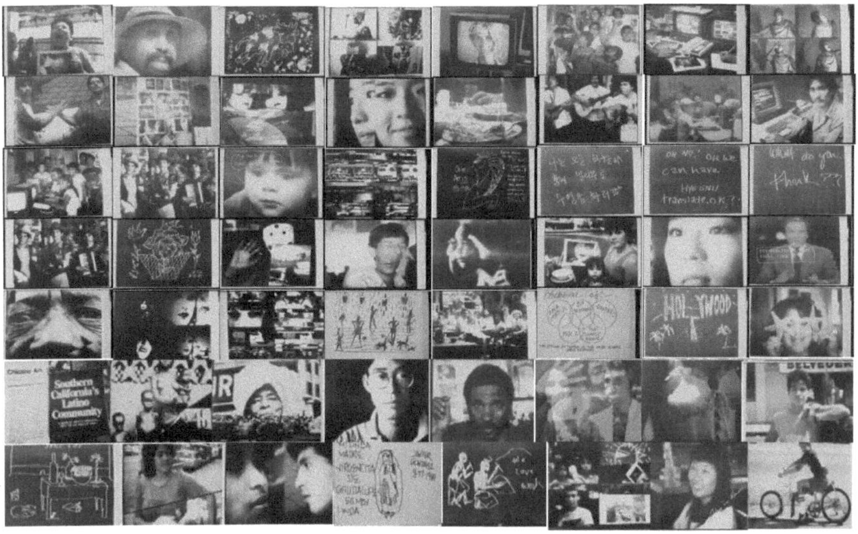

Users at Ana Maria in East LA transmitted music, poetry, and photographs of their community as well as historical images of Mexico, while gangs exchanged pictures of their signs and handshakes, collaging entire graffitied walls from image printouts. Once familiar with the technological arrangement and their place within it, users increasingly reprocessed transmissions, producing ever-more-complex communications. Original screenshots were repeatedly enriched—juxtaposed, superimposed, reassembled, drawn on, adorned with written commentaries—and then sent back out across the network (figure 3.16). The slow-scan cameras, drawing tablets, and optical disk system allowed imagery to be shared among individuals who did not speak a common language. Real-time, digitally produced photo- and text-based montage became a primary language of exchange and collaboration across sites, encouraging users to scrutinize

3.16

Mobile Image, *Electronic Café*, printouts, 1984. Courtesy Rabinowitz/Galloway Archive.

given categories of similarity and difference as well as their relation to a shared apparatus of experience. In one instance, white feminist poets at Günter's in Venice Beach set up a virtual poetry slam, intended as an "overture" to Black male poets at the Gumbo House in South LA.[41] But the exchange remained one-sided, frustrating the Venice contingent's expectations of amicable reciprocity: when the men read, the women responded, but the latter's contributions elicited no reply. The feminist poets then confronted their counterparts, asking them to reflect on their behavior and the gender roles reinscribed by it. When the men tried to rationalize their response (or lack thereof), it set off a heated discussion. These participants engaged the telecommunication technology as a tool of struggle between contradictory points of view that would be resolved not because of the magic of the machines at hand but rather because people could use them to negotiate conflict actively, recognize mutually restrictive categories of experience (in this case, overdetermined identity categories), and eventually find common ground and shared potentialities. In the end, the two groups met in person for a joint reading at Café Cultural, a poetry venue in East LA. As Galloway put it, "There was this whole idea about lack of encounter. LA is a bunch of cities looking for a place . . . It's about encounter; [*Electronic Café*] could support communities—that was the thing—communities not defined by geography."[42]

This focus on community and communication had particular resonance in the regional art milieu of the time. By the mid-1970s, there was a growing sense that Southern California had developed over the previous couple of decades into an important art center deserving of recognition and more substantial institutional support. In 1974, Peter Plagens published the landmark *Sunshine Muse: Art on the West Coast, 1945–1970*, which included a substantial overview of the Los Angeles area. That same year, the Los Angeles Institute of Contemporary Art (LAICA) opened, "designed," as director Bob Smith explained, "to meet the needs of the dynamic, growing, productive atmosphere now identified with Southern California art."[43] Alongside such attempts to bolster the region's self-identity and reputation, however, there was a palpable sense that its art community was still very much a work in progress, with serious questions about its stability and viability. LA's was a peripatetic scene, in which, as Jane McFadden recently put it, "communities would form, for an evening or a decade, in a room, in a magazine, or on the radio, and then dissolve."[44] Uncertainty pervades the early issues of the *LAICA Journal*, which was launched in advance of the Institute's opening and became an important forum for the expression and exchange of ideas during this time. Artists, critics, and curators repeatedly mention Southern California's relatively weak art community, the diffusing effects of the city's sprawl, and its tenuous connections to the rest of the art world, with an almost obsessive focus on LA's perceived provincialism and how the city stacked up against, and could ever hope to compete with, New York. In 1976, Richard Armstrong, then-curator of the La Jolla Museum of Contemporary Art, noted a "general instability" of the art community, attributed to "an essential difference between the east and the west coasts as I know them—between Manhattan and Southern California: cultural surplus. New York has it. Southern California hasn't."[45] For Armstrong, this meant a dearth of contemporary art spaces and publications, which threatened the long-term prospects of a cohesive (and marketable) scene. Overall, there emerged an acute self-consciousness of place and a distinct ambivalence toward it, manifest in recurrent discussions about whether working in and around Los Angeles was an obstacle or whether the qualities of being there—a certain sense of openness and experimentalism, a better climate and more easygoing lifestyle, the lack of centralization, the fact that it was *not* New York—were really the scene's most vital assets.[46]

Most prominent was the discourse around community and the challenges of sustaining it. LAICA and other grassroots institutions such as the artist-founded Los Angeles Contemporary Exhibitions (LACE; 1978) were motivated not to expose people to international trends in contemporary art but rather "to supplement large deficiencies" in the local art community, to be "regional" in both content and purpose. As Smith put it on the eve of LAICA's inauguration, more important than its exhibitions and journal was its mission to bring people together, to become "an essential and unifying catalyst . . . [that] fosters a new atmosphere of communication and cooperation."[47] This was understood as a response to a problem that had plagued the Southern California art community for some time. In a 1978 article titled "Patterns in the Support Structure for California Art," LA-based critic Peter Frank summed it up this way:

> LA (need I remind you?) is a big, sprawling, spread-out region where people tend to congregate only in freeway pileups. Communication tends either to the haphazard or to the ritualized, depending on a deliberate attempt to seek out someone. Except for openings, there is no place one can go and dependably find birds of a feather; there are no real "artists' bars." With the whole situation depending on a high-powered meeting of money and minds that just was not coalescing, Los Angeles art crept into this decade like a mugging victim: battered, impoverished, scared, and disgusted. The final blow was dealt by the recession that crested around 1973, knocking the remaining wind out of the gallery commerce by doing in several regional industries (notably the aerospace industry).[48]

The pockets of art communities that did exist at this time were largely separated by geographic location, which, due to the segregated makeup of the city, also meant division along racial, ethnic, and class lines. And although a hotbed for feminist art and activism during this time, the scene felt no less divided along traditional gender lines. In a 1976 interview between curator Marcia Tucker and LA-based artist Pat Steir, they agree that, in contrast to New York, "in Southern California, the male and female communities are completely split."[49] The few more established art institutions struggled to bridge these divides, faced as they were with economic precarity and increasing criticisms about their conservatism and lack of diversity.[50] The question of how to bring people together across the vast urban sprawl seemed almost existential.

Although part of a national trend, the rise of arts-based collectives and "alternative spaces" in the 1970s therefore took on particular significance and urgency in the Southland. What began in the late 1960s and early 1970s with artist-run spaces such as Continuum (1967), Gallery 32 (1968), and F Space Gallery (1971) gained steam during the ensuing years as groups across the city sought to build community, primarily around identity-based categories. These endeavors were almost always oppositional; as Daniel Widener points out, in Los Angeles, "the creation of alternative cultural institutions and the cultural critique of collectively organized artists offered a popular challenge to the prevailing status quo."[51] Important feminist collectives, including the Feminist Art Workers, Ariadne, Mother Art, Sisters of Survival, Double X, and the Waitresses, emerged, as did the Latinx organizations such as Self-Help Graphics & Art and the Social and

Public Art Resource Center (SPARC) and Black collectives and art spaces such as Studio Z and the Brockman Gallery—the latter founded in 1967 and becoming, by the early 1970s, the center of the Black art community in Los Angeles.[52] As Linda Frye Burnham explains, by the mid-1970s, "artist run spaces were cropping up all over Southern California," mainly exhibiting in public spaces, on the streets and in local establishments, from coffee shops to laundromats.[53]

One of the defining features of the 1970s Los Angeles art scene was its redefinition of what constitutes an art space—arguably also a response to a decentralized scene spread out over the vastness of the city. In her essay for the 2011–2012 exhibition "Pacific Standard Time: Los Angeles Art 1945–1980," McFadden notes that LA artists at this time often exchanged the gallery and studio for what she calls "sitelessness," meaning that "works occur[ed] in a variety of places, both physical and virtual, that were outside the traditional venues of art."[54] The accuracy of that term may be debatable, since so much of that activity was decidedly site specific, explicitly incorporating its place and context, but Los Angeles was indeed increasingly characterized by "a new kind of venue for art . . . consist[ing] of a tenuous organizational network of actions and installations."[55] Frank noted this in 1978, explaining that "there's one form of alternative space that is peculiarly LA: the alternative-space-without-walls, an alternative host structure with no fixed space."[56] Several organizations emerged to support this activity, including Carp (1975) and Some Serious Business (SSB; 1976), both of which held performances and exhibitions throughout the city, and artist-run publications and other distributive media such as records and audio cassettes of sound works were also understood as part of this movement. While the use of electronic and mass media was certainly not unique to the Los Angeles art scene, it was particularly conducive to a diffuse arts community lacking opportunities to gather physically—and in a city so dominated by movie and television industries. Southern California hence became, as Frank observed, "one of the densest areas for the dissemination of such material."[57] McFadden likewise acknowledges the relative abundance of artists there who "experimented with, mined, and struggled with virtual sites of cultural experience, such as radio, television, video, and print journalism."[58] Video was especially important, "serv[ing] as a crucial tool for exploring pedagogy, collaboration, and communication in multiple forms. Video was a medium of community"—especially identity-based community.[59] In her 2014 essay on the queer West Hollywood video theater EZTV, Julia Bryan-Wilson similarly describes this period as "a specific moment for the potential of video production as a tool of community organizing in Southern California," a "key to 'social art'—an early forerunner to what people now call 'social practice.'"[60]

The proliferation of collectives, alternative spaces, and new media activities in the Los Angeles area was thus both highly generative and, at least in part, a sign of and response to the city's long-standing challenges with community building. And whereas experimental solutions emerged, the difficulties of connecting diverse groups and activities were rarely overcome. To Widener, the especially deep racial divisions in the LA art scene were bound up with these difficulties, rooted not only in the country's entrenched history of segregation but also in the unusual physical makeup of the region. In contrast to cities such as San Francisco, where certain postwar figures seemed capable of at least partially uniting different communities, "the infamous Southern California sprawl helped ensure the existence of parallel, and often unconnected, avant-gardes," Widener explains. Unable to link the various art enclaves emerging across the city firmly, "Los Angeles facilitated the development of individuals such as Horace Tapscott, John Outterbridge, and Jayne Cortez, whose artistic production, political organizing, and teaching energy were aimed squarely at black Angelenos and took place largely, though never exclusively, beyond the attention, interest, or presence of politicized white experimentalists."[61] The drive to build artistic community in Southern California brought people together in support of common goals while further dividing up the art scene overall.

A similar situation emerged with the region's art schools and university art departments, which served crucial roles at this time as some of the few physical and discursive "centers" in a distinctively decentralized city. "Of enormous importance," Burnham explains, "were the contributors of colleges and universities, which showed and encouraged this young generation of artists, particularly University of California, San Diego (UCSD), the Claremont Colleges, Otis Art Institute, and the feminist art program at CalArts."[62] At the time, commentators saw these schools as comprising a vital system of support, at least partly necessitated by the early 1970s economic recession but ultimately more suited to the region's needs. As Frank pointed out in 1978:

> For an environment so seemingly hostile to the emergence and growth of new forms southern California has done remarkably well by its artists in the last several years. The main factor in this resurgence of activity has not been private galleries, nor museums, nor the initiative of critics or collectors. It has been the presence of a widespread college and university system that has, in Allan Kaprow's words, acted as the "principal patron" for the southern California scene. The schools, both private and public, have proved remarkably receptive to the creation of whole new formats, new divisions in their curricula, devoted to essentially experimental art research. This, and the general responsiveness to the idea of artists on campus from which it springs, has provided life- and art-sustaining employment for many of the Southland's artists . . . The schools have rushed in where the galleries and museums have feared to tread.[63]

And yet, for the most part, the schools cultivated their own independent communities and identities. Southern California remained "regional" in the sense that, as Doris Cypis explains, "artists and art students tended to move in tribal packs, depending on the school they taught or studied at. Rarely did the tribes connect or exchange, except at art exhibitions at the few and far between venues." The coexistence of disparate art-school groups further fragmented an art scene riddled by the identity-based divisions described by Widener and others:

> Aesthetic and cultural differences were also evident between arts organizations, as each worked almost exclusively within their own racial and gender contexts. Social and Public Art Resource Center (SPARC) focused on Chicano sociopolitical issues; Self-Help Graphics & Art worked primarily with Latino and Chicano artists; the Woman's Building included only feminist women artists. Watts Community Arts Center was mostly African-American; there was an Asian-American film collective and Korean art groups, but there was barely any interaction between them. Meanwhile LAICA, Los Angeles Contemporary Art Exhibitions (LACE) (although initiated by diverse artists, including Chicano artists), Beyond Baroque, and Foundation for Art Resources (FAR) were essentially white. Gender, sexuality, and race were often segregated reflecting the dominant cultural context of the time.[64]

Electronic Café can be understood as both indicative of this broader Southern California art context of connection and separation, community and fragmentation, site-specificity and "sitelessness," and as exceptional in its attempt to explicitly engage, and potentially bridge, those very divisions. Participants posted stories, proverbs, and their perspectives on topics ranging from the plight of South American refugees and culturally entrenched sexism and racism, poverty, and lack of education to schoolchildren wanting to find pen pals in other communities across Los Angeles. These posts often built on one another, transgressing boundaries, forging connections, and articulating common causes. Encountering a system of communication that was emphatically transparent, not necessarily in terms of its physical configuration but with respect to its ideological formation—its organizing categories, its parameters of knowledge and identity, and the kinds of relations it ultimately produced—users recognized and commented on its political potential. Some spoke of it as a more productive alternative to the frustrating state of mainstream television and radio. As one Community Memory post—titled "The Gilligan's Island Syndrome"—explained, "The main limitation to communications technology is and always will be the content of the programming. Gilligan's Island transmitted by direct-broadcast-satellite is still trash."[65] Indeed, it was clear to many that the forms of spectatorship that were rapidly emerging at this time—video playback, satellite, cable systems—may have offered more variety, but access to devices alone did not yield social progress. Another post summed up the situation: "I think the ELECTRONIC CAFE is a wonderful opportunity for the community(s) to define what we want of a communication system . . . It is clear that the corporations already have this technology at their disposal . . . It is now time for all of us to determine our own future by thinking PRACTICALLY about what kinds of uses and creations we can use it for. I like the idea that this is in cafes all over the city because at the very least that is what we all have in common."[66]

Pronouncements such as these exemplify the way *Electronic Café* both facilitated a critique of techno-utopian visions and modeled a productive fantasy that eschewed fanciful hopes for harmonious resolution in favor of a concrete, open-ended process of reflexive interaction and exchange. The project was less successful when it reproduced conventional attitudes of emancipation—attitudes at least partially enabled by the artists' hands-off approach—as a technical rather than political question. For some, the setup implied that social transformation would result from access alone, from "more complete participation in 'community' than otherwise might be possible" as one user enthused. Another post proclaimed that new technologies would become "transformational media" when put into the hands of people broadcasting "messages and examples of LOVE, PEACE AND POSITIVE POSSIBILITIES FOR A HARMONIOUS WORLD."[67] In contrast, the transformative potential of *Electronic Café* emerged at those moments when it conveyed—and turned into a mobilizing impulse—the discrepancy between communication media's means of production (material and intellectual) and the relations of users both to those means and to one another. Only then could participants recognize the limitations of technological protocols and begin to imagine and forge provisional connections that defied predetermined categories. The network became the technology of a popular struggle that demanded the negotiation of the ideal and the real, the ideological and the material, the dominant and the alternative, not as opposites but rather as components of a relational field that was itself a subject of contention. Rather than propose new monolithic cultures of authenticity, *Electronic Café* modeled the construction of overlapping, polymorphous publics, all of which deviated from entrenched conventions of identity and experience.[68] Future imaginings were linked to an awareness of current conditions and their contexts. Participants could thus envision a future based on shared material circumstances and concerns and contrast such visions with pervasive myths of techno-social progress. Fantasy became the generator of truly innovative communication with the potential for real social transformation.

TECHNO-ECONOMIES OF A WORLD CITY

These fantastic moments of critical (and self-critical) communication were especially meaningful within the work's specific urban context. Along with the more general dynamics described by Manuel Castells, M. Christine Boyer, Benjamin Bratton, and other urbanists discussed in the previous chapter, a distinct and increasingly dominant set of political and economic powers were converging upon Los Angeles at this time. The uneven development detailed in Mike Davis's formative study, *City of Quartz: Excavating the Future in Los Angeles* (1990), was a result of this convergence. LA in the 1980s was in the throes of what Roger Keil calls "internationalization," a concerted attempt to transform its identity from regional to global center. Envisioning the city as a prime link to an emergent Pacific Rim economy, Mayor Tom Bradley (1973–1993) championed "Project World City" as the centerpiece of his administration.[69] This involved an alignment of neoliberal policy, development initiatives, and global flows of capital, causing massive reorganizations of space and redistributions of resources, along with a refashioning of the city's image as a multicultural melting pot. The locals and locales connected by *Electronic Café* were already entangled in a worldwide economic, political, and technological network; the struggles with and within this network permeated the very relations that comprised the work's "telecollaborative" apparatus. Situating its explicitly regional and identity-based experiences within an expansive media ecosystem—broadcast news, Olympic spectacle, the presidential election, the high-tech communication devices themselves—*Electronic Café* underscored these entanglements. The work exemplified the fact that the urban milieu was a function of much larger national and international forces.

Globalization plays itself out in the urban political sphere, which is where world cities actually take shape. "Global" and "local," Keil explains, "should not be viewed in metaphors of confrontations between static poles but rather as a process of mutual definition as a result of material relations of power. Not only do traditional communities take up the fight against intrusive global capital, but these communities are restructured in the process, *concurrently changing their political and social realities*."[70] Writing in 1998, Keil notes that this "process of mutual definition" had been foundational in Los Angeles for two decades, with the late 1970s and early 1980s as the key moment in which a fantasy of LA as quintessentially internationalized took hold. This required an expansion of the essential myth of the city as a pure, white haven of sunshine and health, fame and fortune, a myth that had already long obscured the violent histories of people of color and the working classes, of the urban masses who actually built Los Angeles and who would now shoulder the onerous social and economic realities of world-city formation. As Daniel Widener notes, while economic expansion of the Los Angeles area in the 1970s was often heralded as a success story of American capitalism, "from the point of view of working people, 'success' meant outmigration and suburban marginality or defense employment for whites, economic dislocation and social upheaval for blacks, and the proliferation of low-wage, non-union jobs for a mostly Latino and Asian immigrant workforce."[71] Nonetheless, by the late 1970s and early 1980s, a new, similarly whitewashed view of globalization had emerged: what was formerly seen as a mixed blessing, as an engine of growth that also came with challenges of increased immigration and other urban "problems," was eclipsed by solely positive image of organized, beneficial change, of LA as a "fertile matrix of future development," programmatically devoid of people of color and "deproletarianized."[72]

The push for internationalization pushed cultural politics—particularly identity and community based—to the fore, both among LA's neighborhoods and between the local and the global. "Without a constant negotiation of boundaries," Keil writes, "the diverse world city communities that live in close proximity one another would not be sustainable even for a short time." In addition, the cultural politics of this emergent world city tended to divide the urban polity into two camps: the globalists, promoting bourgeoise spectacle and "world class culture," and community culture as a different kind of world culture, based on immigrant and other kinds of ethnic or racial minority experiences. The result was a "constant need to strike new territorial compromises between the local and the global, between spaces used as modules of the global economy and places used predominantly for the reproduction of local community."[73]

This is precisely what was taking place in the types of neighborhoods linked together by *Electronic Café*, particularly among the largely working-class minority communities surrounding the Downtown business district, the newly planned economic and symbolic hub of world-city development. Koreatown, for instance, epitomized global–local interconnectedness, as immigrant entrepreneurs there were completing an international economic cycle that started decades earlier when the United States exported capital, technology, and military power to South Korea. Only officially named in 1980 after a decade of rapid immigration, the neighborhood exemplified the notion that, as Keil puts it, "world city formation is the urbanization of global restructuring."[74] This mostly took the form of Korean small businesses serving low-income, non-white populations typically overlooked and underserved by larger corporations.[75] Filling this niche came, however, with a range of complex local challenges. Sandwiched between the emergent Downtown and what was often cast as an irredeemable "ghetto" of South LA, Koreatown's relative economic stability, improved public education, and decreased crime seemed to confirm the long-standing myth of Asians as a "model minority." In reality, though, residents typically worked longer hours for less pay than others, while receiving inadequate social and health-care services. The community also faced anti-Asian resentment on multiple fronts: non-Korean residents, who still made up more than 90 percent of the newly branded neighborhood; the adjacent Black and Latinx neighborhoods, where many Korean-owned businesses were located; and the city at large, where rising Asian immigrant populations and economic success were often viewed as threats.[76] As sociologists Ivan Light and Edna Bonacich documented in their 1988 study of Koreatown, this led to government interventions that inhibited Korean business growth, including policy changes and the discriminatory enforcement of preexisting regulations. (Such uneven enforcement notoriously did not extend to the Black neighborhood, despite regular resident complaints, which only heightened the resentment between the groups.) In response, the community turned inward, intensifying and consolidating their own relations and forming a largely closed off micro-economy built on what Light and Bonacich call "ethnic entrepreneurship," which "foster[ed] cross-class ethnic solidarities instead of cross-ethnic class solidarities." Koreatown's urbanization of the global at once enhanced the diversity of Los Angeles and further divided its working class along ethnic lines, "leaving unchecked the worst consequences of capitalist social relations."[77]

LA's Mexican population had long struggled with the city's continuous growth and restructuring. As Raúl H. Villa explains, since its nineteenth-century pueblo origins, "the city's working-class *mexicano* population has had an ironic place within this historical metamorphosis, being simultaneously in the geographic center *and* the economic margins of the city. Stated differently, their productive labors have always been essential to the city's growth while at the same time their places of reproduction have been in the way of its ceaseless redevelopment."[78] This resulted in repeated cycles of displacement and "barrioization," the forced physical and social segregation by the repressive forces of dominant culture—allyships of government initiatives, the police judicial system, urban planners, the mainstream media, and private capital.[79] In response, LA's Chicano community of the first half of the twentieth century established a robust "alternative public sphere," comprising community-based magazines and newspapers, customary gatherings in local establishments to discuss pressing issues, an active street culture, and the increased availability of popular commercial media such as recorded music, fostering a sense of cultural unity and opportunities for resistance and activism.[80] As mentioned in chapter 2, the "expressway generation" that came of age in the 1970s and 1980s had been subject to much more aggressive and larger-scale social and physical displacements, as monumental highway construction and urban "renewal" projects of the previous decades violently disrupted the barrio communities of East LA. "These signal developments of the 1950s and 1960s," Villa explains, "materially facilitated Los Angeles's next transformation—into the nation's super-city—and symbolically represented this image to the outside world."[81] In the early 1980s, residents of East LA were acutely aware of this history of displacement and the community's powerlessness in defining their own future. In his 1983 history of the neighborhood, Ricardo Romo captures this sense of disempowerment:

> On the east side, where there is a serious housing shortage and many of the existing houses are of substandard construction, where the schools are overcrowded and the children often poorly educated, where traffic congestion has become a way of life, where smog alerts are still too common, and where residents have but little command over their economic and political destiny, the problems of urban growth are ever present in residents' minds. Lacking ethnic political representation in city and county elected offices, eastsiders have had difficulty in presenting their views on the type of community that their children will inherit.[82]

In fact, the private developers of the adjacent Downtown district dominated state, city, and county politics, thus controlling the future of East LA, from infrastructure to housing to its labor pool.[83] In his 1984 history, Rodolfo Acuña similarly notes that residents were facing the harsh reality that the scourge of "urban renewal" projects would inevitably continue. Acuña ties that reality to both the local, commercially focused "ruling elite" and looming macroeconomic developments that seemed likely to exacerbate the community's plight:

> The Eastside problems of unemployment, gangs, and inferior schooling are phenomena created by the system, not by Mexican Americans. These problems form a very real part of the capitalist system . . . Future changes in production will, in great part, determine the fate of East Los Angeles, generating further alterations in the utility and value of property. The current conversion from an industrially based economy to high-tech production could have disastrous consequences . . . In all events, the future of East Los Angeles is now at a crossroads. The prospects of Mexicans continuing to occupy this community as it is presently constituted look dismal.[84]

The neighborhood's historical "alternative public sphere" would continue to provide essential resources to residents while serving as a nexus of communal "battle against the bulldozers" resistance. Yet, it was simply no match for urban ruling powers fixated on world-city formation.

The contradictions of such formation were arguably most extreme in the African American community of South LA. Long disadvantaged, this community now confronted the irony of a Black mayor championing a vision of Los Angeles as a global center of cultural diversity and interconnectedness while presiding over policies that further impoverished and marginalized local communities of color. The late 1970s witnessed a historic "tax revolt," culminating in the 1978 watershed passage of Proposition 13, a voter initiative that greatly reduced property taxes, causing a precipitous decline in municipal revenue. From the start, this revolt was waged along racial lines, promoted as a way to keep "inner-city" populations from invading suburbia. As Davis recounts: "In rousing their neighbors, tax protestors frequently resorted to the inflammatory image of the family homestead taxed to extinction in order to finance the integration of public education and other social programs obnoxious to white suburbanites." Promoters of Prop 13 also allied their cause with resistance to school integration and busing. The result was "one of the largest mass windfalls of wealth in history" that devastated social services and public schools in South LA, East LA, and other communities of color, while reinforcing their segregation.[85]

The march toward world-city status thus coincided with the entrenchment of the image of Black and Latinx neighborhoods as barbaric, gang-infested, poverty-stricken ghettos. Whereas violent crime did surge in these areas at the time, as Davis explains, it was often exaggerated by the media, feeding "a voyeuristic titillation to white suburbanites devouring lurid imagery in their newspapers or on television."[86] This situation led to increasingly brutal and militarized police crackdowns in South LA, the virtual exclusion of Black people from the broader public realm, including the city's vast stretches of playgrounds, beaches, and entertainment centers, and justifications for urban redevelopment plans that drained resources from the community and further ghettoized its residents.[87] "Characterized by limited transparency, open soliciting of international capital, and the participation of self-appointed groups of powerful citizens," Widener explains, "the cultural dimensions of efforts to build a 'world city' served as a de facto transfer of resources away from South Los Angeles."[88] As Davis points out, however, one very particular global industry—the drug trade—remained available to the community, yet another effect of international shifts in production and capital. "Through 'crack,'" he explains, "[drug gangs] have discovered a vocation for the ghetto in LA's new 'world city' economy."[89]

Mayor Bradley's seeming indifference to the plight of the South LA community, even as he touted the benefits of a multiethnic, tolerant city, reflected a distinct shift in minority-based cultural politics in Los Angeles at this time, one that fit within the broader vision of world-city development. Defined by what Widener calls "the twin imperatives of inclusion and containment," such politics effectively split along class lines in two contradictory directions: on the one hand, a middle-class focus on the urban public sphere as a site of celebratory diversity and community cohesion, championed by a rapidly rising cohort of Black politicians eager to move away from the radicalism of the previous era; on the other, the working-class reality of economic dislocation, political marginalization, and social conflict, in response to which mere survival became the primary concern.[90] Public affirmations of multiculturalism and accompanying municipal policies masked the unrelenting impoverishment of LA's minority communities and the often ruthless efforts to contain them while helping to uphold an image of urban peace necessary for the rise of the global city.[91] As Widener explains, "Inclusion and exclusion formed part of a dialectical exercise of power that functioned as an intermittent reinforcement capable of deflecting popular concerns while allowing unabated upward transfer of cultural and financial capital."[92]

This dialectic of inclusion and exclusion operated within a larger system of "trickle up" development aligned with global capital flows and sustained by public displays of both local cultural diversity and world-class cultural sophistication. "Ethnic" street fairs and other varieties of what Widener calls a "sealed variety of multiculturalism" proliferated at this time, reinforcing a reductive notion of identity-based authenticity, despite histories of multiracial and multiethnic communities in Los Angeles and the intersectional evolution of the city's culture.[93] Large-scale projects were, in turn, designed to "deregionalize" LA, as cosmopolitan culture was elevated over community culture.[94] Grand spectacles, most notably the 1984 Olympics, were to signal the triumphant entry of LA onto the world stage. The arts, though, would arguably serve an even more enduring role in mediating the determined economic processes of internationalization. While local arts organizations, community workshops, and venues such as the Watts Towers Arts Center, the Inner City Cultural Center, and the Bilingual Foundation for the Arts were being drained of public resources, huge investments in large outdoor sculptures, new museums and performance centers, and other public arts programs—the decidedly internationalist Olympic Arts Festival being just one temporary example—protected municipal authorities from popular complaints about inner-city cultural depression, while facilitating corporate redevelopment and providing the symbolic capital fitting to a global city.[95]

This "conjuncture of arts bonanza and scorched earth," as Davis puts it, was nowhere more apparent in the redevelopment of the once-distinct Bunker Hill neighborhood into a gleaming Downtown business district, the symbolic centerpiece of which would be a new cutting-edge art museum. Conceived by Mayor Bradley and several arts patrons in 1979, the museum was imagined as the nucleus of large cultural complex akin to New York's Lincoln Center.[96] Initially called the Los Angeles Museum of Modern Art, the planners quickly renamed it the Museum of

Contemporary Art (MOCA) to signal an international rather than regional focus.[97] The new museum would be designed by Japanese architect Arata Isozaki, and its opening was intended to coincide with the 1984 Olympics. (It did not actually open until 1986.) In 1983, the city opened the Frank Gehry–designed "Temporary Contemporary" as an interim exhibition space—which served, a year later, as the *Electronic Café* node. Almost immediately, the museum loaded itself up with works by blue-chip artists, most notably by purchasing works by Mark Rothko, Robert Rauschenberg, Roy Lichtenstein, Franz Kline, Claes Oldenburg, and James Rosenquist through a "sweetheart" deal with one of its board members, Giuseppe Panza di Biumo, a count from Milan, Italy.[98]

MOCA was understood as essential to the district's development, which relied on an orchestrated, multilayered process of "pushing" out undesirables while "pulling" in desirables. The former involved what Widener calls "new technologies of exclusion," including militaristic police tactics and the destruction and reconstruction of public space, that would effectively drive non-white populations to the margins. World-class institutions such as MOCA could then anchor the "pull," signaling a reclamation of the area "as a public space for affluent Angelinos generally uninterested in traveling downtown after dark."[99] As Jo-Anne Berelowitz explained in a 1990 study: "Clearly intended as more than merely a showcase for art, more than merely the signifier of its own function, MOCA serves also as climate creator for international finance; as catalyst for developing a 'real' downtown; as gathering place; a generator of intriguing experimentalism; a social adventure; a demarcator of innovation; engenderer of honor, attention, business and jobs; monument; destination; and LA's first step toward urbanism."[100] Widener likewise sees the museum as an example of how the arts can be instrumentalized on behalf of "trickle up" development: "The story of MOCA, downtown redevelopment, and the rapid rise of a real-estate-driven 'high-culture' boom in Los Angeles of the 1980s offers a salient window into the place of expressive culture in the exercise of local power. Constituting a kind of cultural revolution from above, the arts-centered downtown redevelopment served as an impetus for the transfer of vast sums, the enactment of new patterns of spatial separation under the aegis of intensely aggressive policing, and a shift in cultural resources from underdeveloped areas of the city to its increasingly parasitic center."[101]

MOCA was thus the cornerstone of an extensive public–private partnership that, starting in the 1970s and 1980s, would radically transform the Bunker Hill neighborhood, literally and figuratively flattening it to make way for the rapid rise of skyscrapers, government and other cultural institutions, and upscale residences. The last bastions of affordable housing were condemned as "slums" and destroyed, residents were displaced, the hill was smoothed out, and space itself was privatized—in terms of both ownership and experience. As Anastasia Loukaitou-Sideris and Gail Sansbury describe in their history of the area, "The redevelopment projects created isolated corporate 'monuments' that turned their backs to public streets, replacing the streets with private internal paseos and plazas."[102] Writing in 1995, they observe: "The present streets of Bunker Hill lack the qualities of complexity, diversity, and contextualism that characterized the earlier landscape. The mega-blocks that are now dominating Bunker Hill have been developed as disjointed and fragmented pieces. The episodic nature of such developments prevents them from effectively connecting with the city's urban tissue. The 'inside' private spaces system systematically exclude the 'outside' public environment. High-rise towers turn their backs to the city; corporate plazas are separated from sidewalks by high protective walls; skyways take pedestrians away from the streets; escalators lead to sunken shopping malls and parking structures."[103] The result was a generic American downtown, devoid of any distinctive qualities of the original neighborhood, stripped of its history, and cleansed of its long-standing working-class minority community. "This," Loukaitou-Sideris and Sansbury explain, "combined with the absence of 'urban clues'—older buildings and urban artifacts that relay the history of the site—prevents the visitor to Bunker Hill from being oriented in space and time, and from developing a complete understanding of the area's social and historical context."[104]

The central hub of *Electronic Café*—MOCA's Temporary Contemporary, along with its surrounding Downtown district—was thus the physical and symbolic locus of a citywide network of power and capital that was, at that very moment, seeking to establish itself within a worldwide economic network, which, in turn, was impacting each of the supposedly distinct neighborhoods linked together by the project. Amid the spectacles of the Olympics and its international Arts Festival, participants were called upon to reinject their own site-specific stories and histories into these intertwined networks. The new, globally focused art museum was recast as a repository for these local histories, moments of exchange, and markers of community, place, culture, and identity; it was where the Community Memory database was housed and where things were sent from the other neighborhoods to be printed, displayed on walls, viewed, recorded, stored, and ostensibly preserved. MOCA and its emergent Downtown, the epicenter of world-city formation, were transformed into a site and archive for that which was being actively displaced, marginalized, and erased.

CRITICAL UTOPIA

Along with the humanist ideal of global unity on which every occurrence of the Olympics is built, LA's games relied on a blend of innovation and corporatization that informed everything from the modernistic design of logos, typefaces, and uniforms to the opening ceremonies themselves, capped off by a man flying into the stadium on a Bell Aerosystems jet pack to light the Olympic flame (figure 3.17). Facilitating the two-week extravaganza was a widely publicized telecommunication infrastructure constructed with approximately $50 million in equipment provided by IBM, AT&T, MCI, and Motorola.[105] (IBM was an official sponsor of these Olympics, the first to be paid for entirely by sales of television rights, tickets, and corporate sponsorships rather than public funds.) Comprising an array of high-tech devices—email, voicemail, searchable databases, credential scanners, online bulletin boards—this "revolutionary" telecommunication system was built to support every aspect of the Olympic operations, from managing time and bodies more efficiently (scheduling, food preparation, transportation) to bolstering security procedures, streamlining internal communications, and enabling the mass dissemination of official information.[106] As an InfoWorld article put it at the time, "The 1984 Summer Olympic Games will have all the technological pomp and polish of a NASA Space Shuttle launch . . . This year's games will be saturated with every piece of computerized equipment imaginable."[107] Along with its celebrated Arts Festival, this spectacular display

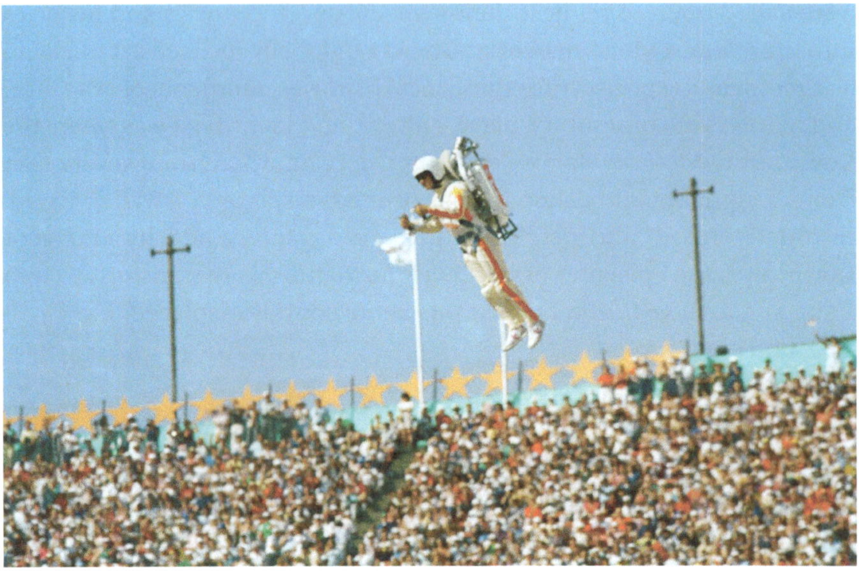

of advanced technology represented a depoliticized dream in which art, sport, and technology work in unison on behalf of humanist ideals, whose future fulfillment was set in some unspecified time and place.

Such future imaginings belonged to the techno-euphoric moment—the "computerphilia" identified by Community Memory a couple of years prior. As mentioned, by the mid-1980s, new telecommunication technologies had become thoroughly mythologized, with the personal computer promoted as the path toward the fulfillment of a utopian vision. Instead of a room-sized machine, the computer became a domesticated tool, deemed capable of freeing individual users from the centralized control of information, labor, and communication.[108] The same year as the Los Angeles

3.17

Bill Suitor flying in on a Bell Aerosystems jet pack to light the Olympic flame, Los Angeles, 1984.

Olympics, Apple unveiled its famous "1984" television commercial, introducing the Macintosh personal computer as a means of liberating users from the forces of conformity and Big Brother—a utopian reimagining of George Orwell's dystopic sci-fi vision.[109] Considered a watershed moment in the history of the personal computer, the commercial contained blatant Cold War overtones, presenting consumer choice as a Manichaean battle between good technology (independent, individualized, democratic) and bad technology (centralized, authoritarian, collectivist, soul crushing).[110] The Macintosh was marketed as the epitome of the former, in contrast to corporate behemoths, namely IBM. Yet, Apple's strategy fit within already prevalent industry bids to humanize the computer. As Ted Friedman recounts in *Electric Dreams: Computers in American Culture*, early 1980s PC ads were focused on "popularizing . . . a new vision of computing as decentralized, democratic, and empowering" as a way to expand the market beyond spreadsheet users and to counter apprehensive views of technology propagated by movies such as *2001: A Space Odyssey*.[111] "Companies bent over backwards," Friedman explains, "to reassure consumers that computers were simple, unthreatening devices."[112] A few years before Apple's Mac ad, IBM launched a campaign featuring the protagonist from Charlie Chaplin's *Modern Times*, but with a decidedly different message. These ads transformed Chaplin's tramp from a nobody caught in a world of hostile technology into a mobile, autonomous, happy worker, liberated by personal computing.[113]

The emancipatory vision of telecommunications was widely embraced as a way for everyone to participate in the making of the future—as an age-old science-fiction dream finally come true.[114] As Turner argues, by the end of the 1980s, "the same machines that had served as the defining devices of cold war technocracy emerged as the symbols of transformation," reconceived as a path to the "counterculture dream of empowered individualism, collaborative community, and spiritual communion."[115] The result was a fanciful "new economy" built on a mix of libertarian politics, techno-utopianism, and counterculture aesthetics.[116] As the history of the "Whole Earth" network makes clear, hippie-era ideals were co-opted by the emerging technological hub of Silicon Valley and the forces of capitalism. Countercultural entrepreneurs such as Stewart Brand championed the conventionally utopian techno-liberationist myths of the day.

Around this time, a number of artists embraced the idea of telecommunication networks as potentially emancipatory tools, a means of resisting media consolidations and institutional constraints. Novel devices were presented as autonomous and available (or soon to be) for free use, anticipating a time when they would be widely accessible and unhindered by the forces of domination. New technology would mobilize and empower artists to operate outside art institutions, wrest control over their tools, transcend physical and geographic limitations, and unify their communities via new means of authentic individual and collective expression. As Marc Ries explains, these projects "expressed a (perhaps diffuse) political will to create the conditions for a social space embracing the *equality*, *participation*, and *accessibility* of and for potentially everyone via technology that genuinely incorporated [a] communitarian ideal."[117] Implicit here was the notion that social and political reorganization—based on collectivism, collaboration, universal access to free-flowing information, and the decentralization and deterritorialization of power structures—could be achieved through a global village of networked individuals freely expressing themselves. Works of art would function as models for this imagined future. Although looking ahead to a world of instant, all-inclusive communication, this vision was largely based on nostalgic ideas of public life and the democratizing function of free exchange, a technologized version of what Jürgen Habermas famously called the "bourgeois public sphere."

British artist Roy Ascott produced a series of works he called "telematic," borrowing a term coined in the 1978 report "*L'informatisation de la société*" by France's Inspector General Simon Nora and Finance Inspector Alain Minc. Nora and Minc described an impending telematic revolution, in which increased interconnection via computers and telecommunications "will alter the entire nervous system of social organization . . . open[ing] radically new horizons."[118] With works such as *Terminal Art* (1980) and *La Plissure du Texte* (1983), Ascott sought "collective consciousness" via disembodiment, dematerialization, and deterritorialization, a liberation of art from the "barriers" of materiality and geography.[119] In "Art and Telematics: Towards a Network Consciousness" (1984), he proclaimed that "computer-mediated networks" open possibilities for "planetary conviviality and creativity" because "networking puts you, in a sense, out

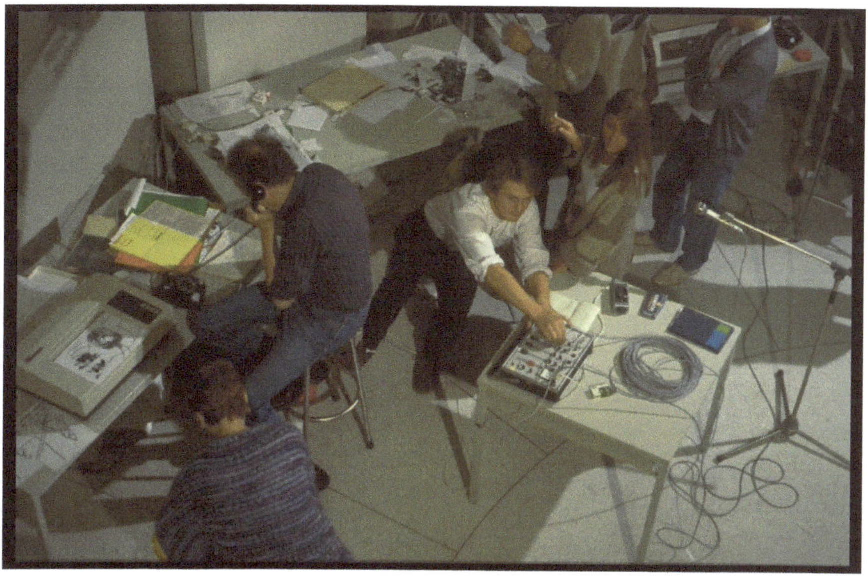

of body, linking your mind into a kind of timeless sea."[120] For Ascott, the technology itself would be transformative.[121]

Other telecommunication projects from this time grappled with this kind of techno-utopianism. *The World in 24 Hours (Die Welt in 24 Stunden)* (1982) was organized by Austrian artist Robert Adrian X as part of that year's Ars Electronical festival (figure 3.18). During a twenty-four-hour period, artists in fifteen locations around the world used an ARTEX-based network of telephones, fax machines, and slow-scan TV devices to transmit works of art—each at noon in their respective time zones—to a central receiving location in Linz, Austria. Adrian understood the project in terms similar to Ascott's, describing it as "a kind of telematic world map"[122] and explaining that it was "intended to develop techniques for

3.18

Robert Adrian X, *The World in 24 Hours (Die Welt in 24 Stunden)*, 1982. Photograph by Sepp Schaffler.

individual, personal, use of existing telecommunications technology" and to "find human meaning in an electronic space."[123] Yet, while enthusiastic about the potential of telecommunication technologies in the hands of individuals, Adrian acknowledged the project's limitations: "Some things didn't get recorded or photographed . . . the sound recording equipment broke down . . . the person with the camera went home to bed, with the camera! . . . we ran out of video tape in the middle of the night when everything was locked . . . one telephone died and another got very neurotic in the early morning . . . we all forgot and lost things, including telephone numbers." He also recognized the disproportionate power relations embedded in the apparatus, "the fact that most of the globe is missing from the network (all of Africa and South America and most of Eastern Europe and Asia)."[124] Some participants expressed ambivalence about the dematerialization and disembodiment promised by telecommunication technology—and celebrated by the very project in which they were participating. In a series of messages transmitted over the network during the event, Eric Gidney (in Sydney, Australia) wondered if the project "IGNORES WHAT WE AS A SOCIAL ANIMAL CONSIDER FUNDAMENTAL NECESSITIES FOR EFFECTIVE COMMUNICATION . . . I.E. PERSONAL CONTACT BODY LANGUAGE, PROXIMITY, JESTURES [*SIC*], PERSONAL IMPRESSIONS ETC. ETC." (Adrian dismissed this query: "IF YOU WANT BODY CONTACT GO TO A MASSAGE PARLOUR.") Gidney also raised questions about gender equality: "IS 'ARS ELECTRONICA' MALE DOMINATED??? THE RELATION OF THE NUMBER OF MEN AND WOMEN IS DOMINATED BY MEN . . . WHERE ARE THE WOMEN??????"[125]

One work included in *The World in 24 Hours* seemed to engage the relationship between disembodied communication and dominant (patriarchal) power relations. *Signal Breakdown—Semaphore Piece*, by Peggy Smith, Nancy Paterson, and Derek Dowden, founders of the Toronto-based space Artculture Resource Center (ARC), paired slow-scan transmissions of Smith using flag semaphore to spell out the words "signal breakdown" with text messages containing emergency warnings of an attack on communication systems by a "feminist art army." *Signal Breakdown* reinjected the body—both literally through the semaphore dance and figuratively through the references to feminist militancy—into a project that generally embraced disembodiment as a path toward a utopian global village. The work suggested that the entangled politics of gender could not be avoided or smoothed over by the network; emphasis on the (female) body complicated Adrian's stated desire to model a unified and universal electronic space, accessibly by everyone and located everywhere and nowhere, for the expression of "human meaning." As Haraway explains, "Feminist embodiment . . . is not about fixed location in a reified body, female or otherwise, but about nodes in fields, inflections in orientations, and responsibility for difference in material-semiotic fields of meaning."[126] *Signal Breakdown* disrupted what Haraway calls the "god trick" of technologically enabled communication, the utopian myth of universal access, objectivity, and equality that "den[ies] the stakes in location, embodiment, and partial perspective."[127]

The precarity of the technological setup in *Electronic Café* was likewise an important part of the overall research project, as the clunkiness and periodic malfunctions of the devices contrasted with their fanciful promise and the sci-fi aesthetic of the consoles in which they were arranged. More generally, Mobile Image attempted to magnify the stakes mentioned by Haraway, producing a work that at once modeled a form of productive future imagining and encouraged critiques of techno-utopian myths. As such, *Electronic Café* also corresponded with a particular discursive shift in America. During the early to mid-1980s, writing on science fiction and the politics of utopia proliferated, in works including, notably, Fredric Jameson's "Progress versus Utopia; Or, Can We Imagine the Future?" (1982), Donna Haraway's "A Manifesto for Cyborgs: Science, Technology, and Socialist Feminism" (1985), and Tom Moylan's *Demand the Impossible: Science Fiction and the Utopian Imagination* (1986), as well as in a special issue of the *Black American Literature Forum* focusing on the work of science-fiction writers Samuel R. Delany and Octavia Butler (1984).[128] Although diverse in their specific subject matter, these texts were united in their attempts to trace the historical development and contemporary pertinence of (and indeed, the urgent need for) future imagining—what Moylan called "images of desire, figures of hope." They all struggled to articulate a truly progressive, innovative engagement with the future that would transform rather than merely repackage the ideological and material limitations of the past and the present.[129]

They also distinguished such an engagement from early modern conceptions of utopia, which both Jameson and Moylan link to the rise of capitalism and its ideological structure. According to Jameson, that structure perpetuated a notion of "progress" that validated enduring discrepancies between everyday experiences and the promises of tomorrow. As Moylan explains, early utopian fictions of the future "provided images of alternatives to the given situation which, while not yet existing in history, drew on the contradictions of the time and anticipated a response to the conflicting needs of the dominant and subordinate classes."[130] Deeply embedded in the colonial and imperial ethos, these "alternatives" were first projected onto the uncharted geographical spaces of the New World and then later relocated to another time—to the future, when truly revolutionary change would ostensibly yield a perfected society.[131] However, as part of a class struggle over the means of material and cultural creative production and exploitation, such utopian visions have the power to both symbolically appease and politically challenge: "Utopian dissatisfaction and imagery has been enlisted into the process of the creation of needs subordinated to the demands of production and profit; while, on the other hand, the very dream-making activity of the utopian imagination continually resists the limitation of human desire to the economic and bureaucratic demands of the given system."[132] As Robert Elliot Fox discussed with regard to Delany's mid-1970s novels, this "dichotomy of experience" of both gratification and alienation accounts for the resurgence of utopian desires in a specifically American context. The political upheaval of the 1960s and its aftermath produced a creative activism, particularly among feminist and Black writers and artists, that sought to raise consciousness about and transcend the commercial-industrial machine of the utopian dreamscape as an engine for the sale of material and immaterial consumer goods, from advertisements for suburban lifestyles to Hollywood values and Disneyworld ideals.[133]

Crucially, this surge in subversive utopianism was self-consciously critical, acknowledging the dialectical nature of utopia, its historical failures, as well as its transgressive potential. What Moylan terms "critical utopia" revives future imagining but refuses predetermined solutions and resolutions, dwelling instead on the conflicts between present and prospective conditions.[134] It "rejects utopia as blueprint while preserving it as dream."[135] For example, Delany's narratives articulate what Jane B. Weedman calls the conflict between "the prevailing idealism of the American dream and Black American reality"; they manifest a "double consciousness" in W. E. B. Du Bois's sense of the term, Weedman explains, "a psychological dichotomy which results when an individual lives in a culture, such as the black community, yet must be aware for his survival of the workings and expectations of a dominate [sic] culture."[136] Rather than paint a beautiful picture of the future that only distracts from the often-violent constraints of the given, a critical utopian practice has as its subject the very construction of utopia, not as an inherently human need but rather as a systemic necessity. Looking to the future is thus understood as a highly contested anticipation of the not yet concrete that determines the relationality between past, present, and future experiences and perspectives, who and what gets to be part of the future imaginary, and whose histories will determine the outcome of what tomorrow looks like.

As a critical utopian project, *Electronic Café* was distinct from various works by other artists and from Mobile Image's previous work. Whereas *Satellite Arts* and *Hole in Space* implicitly politicized telecommunication technology by encouraging participants to think dialectically about the relationships between materiality and dematerialization, territory and deterritorialization, emerging possibilities and imposed constraints, *Electronic Café* explicitly couched such relationships in those "material-semiotic fields of meaning" that Haraway sees as constituting categories of difference. In 1982, Galloway and Rabinowitz taught a performance course at Loyola Marymount University titled "Aesthetic Research in Telecommunications," in which they attempted to reinsert the body emphatically into an ostensibly disembodied technology. Similar to *Satellite Arts*, the course was held in the composite-image screen, exposing students to the reality of "life in virtual space." From multiple sites across campus, participants engaged in everyday tasks, collaborative problem solving, and theatrical skits mimicking real-life situations.[137] In one example, two students interacted "physically" from different locations, not only "touching" each other sexually but also melding together in erotic ways, achieving a level of intimacy, an "electronic foreplay."[138] The course was Galloway and Rabinowitz's attempt to develop a more embodied telecommunication experience. Yet, it was primarily focused on individual expression and interpersonal contact, emphatically private, apparently apolitical, and grounded in the notion of the autonomous, whole (bourgeois) self.

Building on the lessons of this course and their earlier artworks—the lack of a wholly autonomous "third space" in *Satellite Arts* and the fact that communications were encumbered by delays, feedback, and breakdowns, the failure of many *Hole in Space* participants to think beyond a narrow set of behaviors and their own personal desires—*Electronic Café* underscored the specificity of bodies and places, identities and localized perspectives, in relation to the larger technological apparatus and dominant power structures. As Galloway and Rabinowitz explained in their preparatory notes for the project: "Electronic Cafe network is designed and presented as a 'creative solutions network'—creative conversation among/between the divergent cultural communities that make up the greater Los Angeles community. It is designed as a forum for shared ideas/exchange of Art, cultural concerns, collaborative attempts at approaching and proposing solutions to common problems as well as expressing the unique social influences that are brought to the common culture of Los Angeles."[139] Whereas the idea of a "creative solutions network" sounds very similar to the notion of a global electronic village based on collectivism, collaboration, and universal access, an updated version of the bourgeois public sphere, in practice, *Electronic Café* moved beyond that archetype. It embraced new tools while setting into critical relief the typical utopian prophecies of an idealized, electronically networked public of free-flowing individualized expression—still somewhat operational in *Satellite Arts* and *Hole in Space*—along with corporate-driven promises divorced from, or designed to obfuscate, power structures and broad political and ideological realities. It was a model of contentious relationality defined by the conscious transgression of existing geographic, racial, cultural, and technological demarcations through expanded communicative processes. *Electronic Café* at once engaged bourgeois ideals and departed from them, advancing a more self-critical, fragmented, and dialectical notion of the public sphere. As Negt and Kluge explain:

> The bourgeois public sphere is anchored in the formal characteristics of communication; it can be described as a continuous historical progression, insofar as one focuses on the ideas that are concretized within it. But if, by contrast,

one takes its real substance as one's point of departure, it cannot be considered to be unified at all, but rather the aggregate of individual spheres that are only abstractly related. Television, the press, interest groups and political parties, parliament, army, public education, public chairs in the universities, the legal system, the industry of churches are only apparently fused into a general concept of the public sphere. In reality, this general, overriding public sphere runs parallel to these fields as a mere idea, and is exploited by the interests contained within each sphere, especially by the organized interests of the productive sector.[140]

Modeling a technological practice of counterpublicity, *Electronic Café* compelled users to confront their own inscription within existing power structures, enlisting them as producers, receivers, and manipulators of content.

TECHNO-PUBLIC SPHERES

Broadly defined, the public sphere is made up of the instruments and spaces that organize such experiences and perspectives. As discussed in previous chapters, these tools and sites of communication and knowledge production include technical devices such as computers and televisions, as well as materials and institutions such as those listed by Negt and Kluge above. Together, these determine what they call "the social horizon of experience."[141] Any critical-transgressive activism must consider the means of production in this expanded sense as a struggle over the tools that dictate the future imaginary. Because such arrangements of collective experience are subject to power systems and to determined notions and rituals of subjectivity, community, identity, and history being amplified or repressed, they are inevitably bound to apparatuses of publicity, the mechanisms by which people both transmit and receive information, attitudes, and desires. In this sense, "technologies" constitute the public sphere. Thus, a critical utopian practice has to address pervasive myths of technological progressivism.

Galloway and Rabinowitz met in Paris in 1975, each working on expanding the City of Light's perpetual modernity through various electronic-communicative projects. They were introduced by the French philosopher Felix Guattari, himself deeply invested in tele-connective activism. Guattari initially brought Rabinowitz to France on behalf of the French Ministry of Foreign Affairs to show activists there what Americans had been doing with small portable video equipment, and the three remained in touch, later collaborating on a proposed Paris site of the forthcoming *Electronic Café International* (figure 3.19). All three shared an interest in social change and technology and were skeptical that the promise of telecommunicative connectivity would lead to material and intellectual emancipation. According to Guattari, revolution came not simply through more communication—"They talk, oh yes indeed, they talk all the time"—but rather through a specific type of communication, one that destroys "the domination of isolation."[142] Such communication would connect the fragmented communities of the dominated in their desire to overcome alienation and to create an image not of a future society but rather of "a collective competence" through "collective action."[143]

According to Galloway and Rabinowitz, *Electronic Café* was consciously modeled on a particular idea of the French coffeehouse as a site of critical public debate and a cradle of revolutionary action: "This was a new social institution. It took the idea of the café, as an informal human institution, and you could create revolution, and it was distributed. This was from the French model. There was poetry, revolution, all kinds of stuff. It was like the news used to be carried around by troubadours, where they traveled around with mandolins and told you what was going on."[144] For these artists, the function of the new high-tech coffeehouse went beyond traditional accounts of the café's role as part of the generation and transformation of the bourgeois public sphere, the ideal arena of an inclusive, autonomous critical exchange.[145] Deeply immersed at the time in the context of and discussion surrounding communication and power, they recognized that a simple reproduction of then-current communication structures based on a traditional notion of bourgeois publicity—predicated on what Miriam Hansen calls "formal conditions of communication (free association, equal participation, deliberation, polite argument)"—would not suffice.[146] Echoing contemporaneous sci-fi and utopia debates, writers and theorists such as Negt and Kluge, Guattari, Herbert Schiller, and Gene Youngblood cast a critical eye on myths of the democratization of culture through technological access and innovation. Schiller and Youngblood had close ties to the California art and

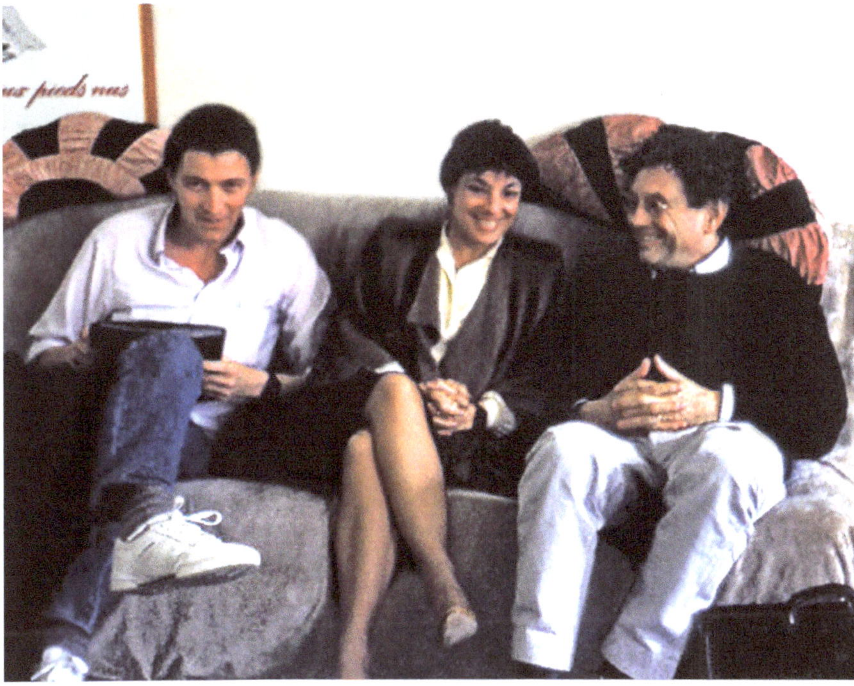

3.19

Kit Galloway, Sherrie Rabinowitz, and Felix Guattari, Paris, 1988. Courtesy Rabinowitz/ Galloway Archive.

technology scene of those years: Schiller was a professor of communication at the UCSD, and Youngblood was a Los Angeles–based critic and professor at the California Institute of the Arts, who would become a close collaborator of Mobile Image. Both theorists took part in *Televiews*, a 1981 video-relay project at UCSD led by the artist Ulysses Jenkins. This project was closely monitored by Galloway and Rabinowitz, who later invited Jenkins to play a pivotal role in *Electronic Café* as an artist-in-residence at the Gumbo House location.

In a lecture as part of *Televiews*, Schiller warned of overly enthusiastic technological forecasts couched in terms of undifferentiated public access and ideals of liberation through technological innovation that promise flexibility, creativity, and control. In his 1976 book *Communication and Cultural Domination*, he had explicitly referred to mass media as "public" media, addressing the ideological dimension of how information is produced and disseminated.[147] If technology is perceived as just a set of devices—hence, as politically and ideologically neutral—it merely serves to relay the same messages produced elsewhere in society. New technology, Schiller explains, does not automatically produce a new society. Myths of ideological neutrality and universal benefit obscure the fact that "free flow" is "a one-way street for exercising domination by the already-powerful, is extended to technology—with the still greater likelihood of intensifying the dependency of the weaker parties."[148]

Like *Televiews*, *Electronic Café* aimed to reveal this dimension of the political utility of communication devices—the fact that, as Schiller put it, "technology is a social construct."[149] Its futuristic consoles pointedly alluded to the facade of commercial techno-progressivism, while each also incorporated a live television feed—including news, entertainment, live events from the Olympic Games, and coverage of that year's presidential campaign—whose imagery could be combined, juxtaposed, written on, and otherwise reprocessed and rebroadcast across the network. Users of the drawing tablets, slow-scan cameras, and other devices thus saw their own creations and manipulations continuously juxtaposed with those of a seemingly omnipotent media industry. Such an encounter with the mechanisms of information production and dissemination is crucial to fighting what Youngblood refers to as "the cultural imperialism of the mass media."[150] Users became producers, appropriating, transforming, and relaying the information and material they received, but the quality of their transmissions contrasted with prevailing modes of distribution based on conventional notions of a monolithic, all-encompassing sphere of idealized public exchange.

Whereas *Electronic Café* permitted people to utilize mass media and telecommunication technologies, the work did not simply provide a way to send out personal commentaries into some preset apparatus or abstract, uniform ether. It was not just a matter of access to broadcasting devices and channels so that people could participate in "free flow." Rather, the nature of their communications was emphatically particularized, directed, and localized, enacted between various interdependent publics from specific café locations and according to continuously negotiated concerns that transcended predetermined categories of identity and experience. As the project made clear to users, the notion of a singular public was replaced by that of multifarious publics that were relational, specific, contested, and grounded in myriad overlapping sociopolitical contexts.

Similar to the German debates between Jürgen Habermas, Negt and Kluge, and other critics that had filled the feuilletons and university seminars since the late 1960s, US discussions regarding who partakes and how in the construction of perspectives, attitudes, and policies were centered around questions of what constitutes a body politic in the context of new social movements, advancements in communication technologies, and increasingly deregulated economies, all engaging in a struggle over greater varieties of seeing and being in the world. As discussed in relation to *Hole in Space*, 1980s writers such as Manuel Castells and Rosalyn Deutsche historicized and politicized the concept of the public sphere, its contemporary function, and art's role within it. This discourse, to which *Electronic Café* relates, inherently involved the politics of participation in the production of knowledge, information, and experience. As feminist, postcolonial, and urbanist critiques sought to politicize the myths of the democratic makings of culture, two crucial notions emerged. The first was that there never was just one single public and thus not only one public sphere, one horizon of experience. Second, if there are multiple publics, how are they constituted and what is the relation between them, between, as Haraway put it, "us" and "them"? What constitutes a "we" in relation to other constituencies, to other bodies, and how can these relationships be emancipatory and avoid the reproduction of the exclusive, hegemonic structures of public and private, center and margin, self and other that are at the core of the logic of neoliberal inequality?

As Mary P. Ryan, Nancy Fraser, and Michael Warner articulated in their respective discussions of the 1989 English-language publication of Habermas's seminal *Structural Transformation of the Public Sphere*, the "political ideal of open, inclusive, and effective deliberation about matters of common and critical concern" was always dependent on *exclusions*, by the demand that in order to act as a *public* citizens would leave *private* matters and concerns at the door.[151] As a consequence, women, whose concerns, bodies, and experiences were consigned to the "domestic," and other groups, whose gender, sexuality, race, ethnicity, religion, or culture were deemed to be "personal" rather than social concerns, were effectively denied access to and representation at those sites and forums where all that is actually or ostensibly relevant to all members of society would be subject to deliberation and, consequently, materialize in the form of official and unofficial policies and resources serving the "public good." In her nineteenth-century history of the US public sphere, Ryan shows that it has been the "imperfect public" of "working men, immigrants, African-Americans, [and] women" that has driven the evolution of institutions and discourses of "American public life" rather than the supposedly autonomous and accessible bourgeois public sphere persistently idealized in theories and practices of democratic agency.[152] Fraser, in turn, argues that multiple, competing publics are desirable precisely because they challenge the purported autonomy that serves the (private) interests of those dominant groups who benefit from a lack of public discussion of questions of ownership, capital, exploitation, and labor. For instance, as Fraser explains, "if questions of workplace democracy are labeled 'economic' or 'managerial' problems and if discourse about these questions is shunted into specialized institutions associated with, say, 'industrial relations' sociology, labor law, and 'management science,' then this serves to perpetuate class (and usually gender and race) dominance and subordination."[153] Contestation, Fraser argues, is the key function of "subaltern counterpublics"—"parallel discursive arenas where members of subordinated social groups invent and circulate counterdiscourses to formulate oppositional interpretations of their identities, interests, and needs."[154] If the ideas and perspectives produced by the dominant institutions and apparatuses of the public sphere are thus necessarily partial yet prescriptive, if they clash with but (in)form the knowledges and experiences of those individual and social subjectivities excluded from officially contributing to the general horizon of social experience, then what is the quality, the politics of the relations between the various spheres and groups, between a contested multiplicity of outlooks, needs, and desires, between "imposed narratives" and "true stories," between constructed and "genuine" images, imaginaries, and corporealities? As Haraway argues, we need to historicize knowledge and critique the technological modes by which meaning is produced while committing to faithful accounts of a "real" world.[155] To assign greater or any singular notion of "objectivity" to one particular knowledge or experience over another runs the risk of playing into the hands of existing power, especially during the Reagan era when "any collective historical subject that dares to resist the stripped-down atomism of Star Wars, hypermarket, postmodern, media-simulated citizenship" was dismissed as a "special-interest group" and thus easily policed in the realms where what does and does not count as relevant knowledge is ascertained.[156] The goal is to overcome the very logic of public and private that is at the core of hegemonic ideals and practices of the traditional public sphere and its institutions and apparatuses of publicity, of mediation and representation, and the ways in which it translates into social and ideological dynamics of "common" knowledge versus embodied experience, of us versus them.

The struggle for what Haraway calls "feminist objectivity [which] means quite simply *situated knowledges*"—hence, the ongoing, historically specific negotiation of individual and social subjecthood—finds its powerful correlate in the critical race and postcolonial debates of the 1980s and 1990s.[157] Authors such as bell hooks, Trinh T. Minh-ha, Gloria Anzaldúa, and Homi Bhabha theorized material and immaterial spaces and positionings that resisted the binary of assimilation and segregation/separation.[158] As hooks put it, "To be in the margin is to be part of the whole but outside the main body."[159] As for Fraser, this positioning on the periphery bore the potential for contestation. hooks framed marginality as "more than a site of deprivation" but one of "radical possibility, a space of resistance": "It offers the possibility of radical perspectives from which to see and create, to imagine alternative, new worlds. This is not a mythic notion of marginality, it comes from lived experience."[160] Yet, as hooks emphasizes, this space does not exist apart from but rather in reciprocally determining relationality to the center and thus, as a site of critical engagement and emancipation, of transformation, does not and cannot insist on/reproduce the mythical autonomy and active depoliticization that defines the exclusionary mechanisms of the bourgeois public sphere: "I want to note that I am not trying to romantically re-inscribe the notion of that space of marginality where the oppressed live apart from their oppressors as 'pure.'"[161] The margin becomes, in turn, a space of radical openness: "As a radical standpoint, perspective, position, 'the politics of location' necessarily calls those of us who would participate in the formation of counter-hegemonic practice to identify the spaces where we begin the process of re-vision . . . For many of us, that movement requires pushing against oppressive boundaries set by race, sex, and class domination, . . . We return to 'identity' and 'culture' for relocation, linked to political practice—identity that is not informed by the narrow cultural nationalism masking the continued fascination with the power of the white hegemonic order."[162] Here, difference, as an oppressive imposition as well as a potential for emancipation, is constructed, historical, relational; the key is to recognize the real as determined by the imagined, and to reimagine and reconstruct the real beyond its existing imaginary confines. As Bhabha elaborates: "What is theoretically innovative, and politically crucial, is the need to think beyond narratives of originary and initial subjectivities and to focus on those moments or processes that are produced in the articulation of cultural differences. These 'in-between' spaces provide the terrain for elaborating strategies of selfhood—singular or communal—that initiate new signs of identity, and innovative sites of collaboration, and contestation, in the act of defining the idea of society itself."[163] These in-between spaces are where public and private, knowledge and experience, center and margin connect and determine one another. They are the sites and moments where, as Negt and Kluge put it, "the real social experiences of human beings, produced in everyday life and work cut across such divisions."[164] The only originality or objectivity—the only horizon there is—is the dialectic of knowledge and experience, of that which ought to be and how the ideas and ideals of our world are being felt, perceived, experienced in particular and overlapping everyday context and circumstances. There is no "god-trick," as Haraway calls it, no outside position to be safely taken; the general horizon of social experience, the public sphere as "the act of defining the idea of society itself," needs to be acknowledged, theorized, and performed as this dialectic of the in-between, of "situated knowledges."

Technology, and communications technology in particular, is crucial to publicity and positionings. In Habermas's account of the bourgeois public sphere's detrimental demise, apparatuses of human connectivity provide the means to transcend the limits of geographic and physical confines, while scientific progress allows for an ever-expanding, rational assessment of the workings of the world. To Haraway, technology provides the opportunity to (re)connect perception to the body, or rather a multiplicity of bodies, to politicize vision and mobility as they are tethered to individual physicalities and social corporealities, to locales and sites, streets and schools, bedrooms and boardrooms, archives and carnivals, and to the tools of transmission and transportation themselves. As such, technology never fully transcends; it generates only partial, never absolute, knowledge or authentic experience. Thus, one connects and is connected only ever in relation to. This is "preferred positioning": "partial, locatable, critical knowledges sustaining the possibilities of webs of connections called solidarity in politics and shared conversations in epistemology."[165]

As opposed to the ostensibly polite and objective (read: disembodied) deliberations of the public sphere, the situated knowledges of counterpublicity demand and depend on engaging the physical presence of its constituencies. And, as Haraway suggests, the body politic(s) of a critical electronic urban public space provide(s) opportunities to connect the corporeal and the immaterial in a progressive, critical utopian manner. In her 2005 assessment of what she calls "the new urban spatiality," sociologist Saskia Sassen discusses the overlaid and interconnected economies of digital and material networks "as cities and urban regions are increasingly traversed by non-local circuits," raising the question, "what is urban place in this context?" The result of this inquiry is a "repositioning of architecture, planning, and urbanism generally, *as forms of knowledge and forms of practice*."[166] The destabilizing effects and the rescaling of "the local" brought about by the "complex imbrications between the digital and the non-digital" favor economic relations of monetary growth and urban development when networks of information and capital are able to move freely, unencumbered by material boundaries and entities. As Sassen explains, "Hyper-mobility and de-materialization are usually seen as mere functions of the new [communications] technologies. This understanding erases the fact that it takes multiple material conditions to achieve this outcome and that it takes social networks not only digital ones."[167] The displacements and labor needed to make these networks viable remain largely invisible, as do the bodies forcibly mobilized and exploited, a condition exacerbated by the aesthetic, economic, and ideological privileging of the digital circuit as inherently progressive. Echoing some of the critiques offered by Manuel Castells and others discussed in chapter 2, Sassen focuses on the complexity of intra- and interurban "networked sub-economies operating partly in actual and partly in globe-spanning digital space" benefitting the flow of capital and goods, forcing a de-territorialized and decontextualized and increasingly alienating reconfiguration of select multiple "locals" no longer defined by traditional boundaries and notions of "neighborhood" but using place as resource where and whenever desirable.[168] Without explicitly articulating it, Sassen's analysis suggests a counterpublic *Umfunktionierung* of the subeconomic model, generating new forms of knowledge and practice made available by the strategic appropriation of technologies typically cast as vehicles of fabled transcendence, but here reconceived as tools for reimagining networks and constellations of bodies, information, territory, and agency.[169]

The suppression of the material and physical from the logic of the new urban spatiality recalls the very foundations of the bourgeois public sphere as described by Peter Stallybrass and Allon White in their 1986 study *The Politics and Poetics of Transgression*.[170] According to the authors, by the late seventeenth century, the European coffeehouse had been established as a space where democratic access and conduct was equated with a particular type of decorum and etiquette, "synthesiz[ing] aspects of both upper-class and protestant morality with respect to clean living and refinement."[171] Its privileging of "de-libidinized" encounters, manners, tone, and behavior—stratified more often than not according to differences in class, gender, race, to bodies and locales, education and socialization—effectively instituted parameters of public control (and control of the public, the body politic). The coffeehouse institutionalized a "protestant ethic," in which the regulation of the "unruly" body is a prerequisite for participation in the public sphere.[172] Following Habermas's account of the transformation of the public sphere via the privatization of public spaces and media, this regulation extends far beyond the sites and occasions designated for rational discourse, beyond deliberation itself, to all spaces of representation and performance that produce and reproduce individual and social subjects and subjectivities, the relations and conduct among selves and bodies. As Sassen explains, technologies of communication and mediation have at once irrevocably destabilized conventional urban relations and reinforced the dominance of economic "centers," now redefined as combinations of "capital fixity and hypermobility."[173] Such technologies extend what Stallybrass and White call the "re-alignment of place, body, status, and discourse" by strategically connecting specific sites, spheres, and "clienteles," while conveniently ignoring and actively marginalizing others.[174]

Certain artists and activists have struggled to re-function these instruments to not only reassert the material and corporeal dimension of visual representation, information exchange, and knowledge production but also—and in light of the much discussed transition from a Fordist to post-Fordist economy, from material to increasingly immaterial labor, including the making of lifestyles as much as the realms of domestic, reproductive, and care labor—imagine and put forward new modes and presences of being. This critical-fantastic assertion of existing and newly created corporealities and networks, or subeconomies, was at the heart of select projects and practices contemporaneous with *Electronic Café*. *The Peoples of Los Angeles*, for example, was also commissioned for the 1984 Olympic Arts Festival. Created by Werkgruppe, a collective founded by artist Daniel Martinez and cultural theorist D. Emily Hicks, the work consisted of nine sculptural forms, each containing a forty-five-second holographic "portrait," exhibited at the USC Atelier gallery. In one, a factory worker was shown dancing with his mother in front of a group of friends and relatives. In another, a man named Jesus wore a rainbow outfit and alternately played a guitar and carved a wooden cross.[175] According to writer Susan Otto, "The artists' plan was to interface cutting-edge image technology with the ancient practice of oral stories . . . Various people of disparate race and class positions retold personal narratives, which were immediately translated into nine languages and intimately heard through suspended headphones." To accomplish this, Werkgruppe developed a "mobile holography unit"—a special camera designed by physicist Lloyd Cross, mounted on a customized wheelchair—that could be brought throughout the city to film people on-site, in their homes, and on their neighborhood streets. Located in the Santa Monica Place shopping mall, USC Atelier was selected because it presented a fashionable "public location for meeting and commerce."[176] This "documentation of people's lives," as Hicks called it, inserted bodies of popular culture into public space, bodies often overlooked or otherwise marginalized in and by its location and the institutions (artistic, commercial, etc.) represented by it. The work at once expanded public space and its constituencies (and clientele) and challenged its logic and utility.

To Martinez, the work was a strategic intrusion into the urban imaginary, part of a practice that sought "direct action and intervention into the landscape, into the aesthetics, into the politics."[177] Like Haraway's cyborgs, these bodies are the "illegitimate offspring of militarism and patriarchal capitalism." Yet, as subjects who refuse to adhere to the confines of modern subjectivity/subjecthood and its reification as Other, they have the potential to resist it.[178] They are both social reality and fiction. These bodies are neither the locus of polite debate in the ideal bourgeois sense nor easily identified and catered to as consumers and hence confined to social, psychological, and corporeal containers and categories of individuality and subjectivity. Thus, they are not easily controlled and surveilled in and through the discrete spaces of public and private, of consumption, work, and home, neighborhoods, peripheries, and centers. Although Haraway's cyborg is an explicitly feminist paradigm, it represents a *politics* of resistance that applies to any bodies, past, present, and future that, as she puts it, "have no truck with . . . seduction to organic wholeness . . . , the 'West's' escalating dominations of abstract individuation, an ultimate self untied at last from all dependency, a man in space."[179] In a similar spirit, Werkgruppe sought to transgress the distinctions between bodies and images, between material and immaterial entities and spaces as discrete, originary sites of defining and managing expectations, attitudes, and experiences. As Otto explains about the work: "The hologram, which can be seen from 360 degrees at all times, allows multiple points of opinion and perspective at once. The holographic narrative is not linear—it mimics a circle, or continuous play. Like the Mexican codex, it allows participants to jump in and out of the narrative at any place. Sound was used as vibration and texture, in a physical sense, in a composition of antitonal structure and movement. It was choreographed to pass from corner to corner throughout the physical space. Layered within this, the machine's motors whirred like insects."[180] The "creatures," to borrow Haraway's term, are not easily confined; they are slippery, partial, multiple, "fluid, being both material and opaque," obvious in their construction and playfully perverse in the performance of their reconstruction.[181] They speak several languages and make literal their objectification as sites of ideological projection. Their voices mix with the sounds of the machine, of "insects": "The boundary between physical and non-physical is very imprecise for us," declares Haraway.[182] The mall, a site for the affirmation of the modern subject through the exercise of public engagement performed and regulated in and through the commodification of participation, choice, and freedom, is turned into a site of carnivalesque defiance. In Sassen's terms, technology provided both a spectacle of the local and the model for a new subeconomy of knowledge and practice through a complex constellation of materially and ideologically expanded bodies and spaces.

A similar politics of transgression, of what Haraway calls "leaky distinctions,"[183] can be found in the work of performance and video artist Ulysses Jenkins, a friend of Martinez's and the official artist-in-residence at *Electronic Café*'s Gumbo House location. Jenkins had worked with Senga Nengundi, Maren Hassinger, Barbara McCullough, Frank Parker, and David Hammons in the late 1970s and early 1980s, creating urban interventions, occupying and transforming places through ritualistic movement, literally and metaphorically struggling to find new forms of being in space. Such work sought to confront and transcend the ways in which the presence of certain bodies in certain spaces had been depoliticized in Roland Barthes's sense, that is, mythologized through habitual seeing, an affirmation of the hegemonic aesthetic logics (and economics) of race, gender, class, and site.[184] For Jenkins, such mythologizations were embroiled in the media and its technologies of representation and dissemination. The artist was concerned with how not merely to substitute one image or myth for another but rather to politicize the given as well as the potential function of bodies in space. "Our work is political," Jenkins explains, "and the politics [are] us being ourselves."[185]

Jenkins's *Dream City* (1981) was both a performance and a video work, comprising a succession of fifteen-minute actions by different artists. As described by the curators of "Now Dig This!: Art and Black Los Angeles 1960–80": "This almost twenty-four-hour event presented to a paying audience involved performers from various backgrounds, including Mexican Americans, Asian Americans, and African Americans, among them fellow artists Maren Hassinger and Senga Nengudi. Multiple characters and objects interact in a fantastical dream state, linked through a story being told aloud."[186] A crucial aspect of the work was the creation of a feedback loop: at the commencement of each participant's performance, video recordings of prior segments were played back. This produced a sense of surveillance while instilling an awareness of a technics of representation and mediation, both circular and as a site for departure and transformation.[187] Communications and imagin(in)g technology, "the reconstruction of how an image is seen, how to produce a new point of view," as Jenkins puts it, was to be a central component of the work.[188] Staged in part as a response to the 1980 election of Ronald Reagan, both the performances and the culminating video articulated a critique of and a desire to surmount the limits of the kinds of fantasies then serving as guides for an urban life to come: the "dream" in *Dream City* refers to the "American dream" as much as it does to the longing for something not yet become. The video shows an assemblage of settings and performing bodies; people dance, read poetry, play music, and enact other activities—from the mundane to the seemingly spiritualistic—some in the studio, others in a concert hall or out in the street. Evocative images bleed in and out: shots of the Los Angeles and New York skylines; an empty corporate boardroom; churning oil pumps. Everything is bathed in technologically enhanced psychedelic colors that give things a glow of artificiality and, like the array of sounds, transgress discrete scenes, blurring boundaries between people and objects and their surroundings. Jenkins himself appears nude except for a Pharaonic headdress and an amulet around his neck (figure 3.20); he is a shamanic figure conjuring both an ancient past and an (Afro)futurist tomorrow. He stands next to a push mower, ready to

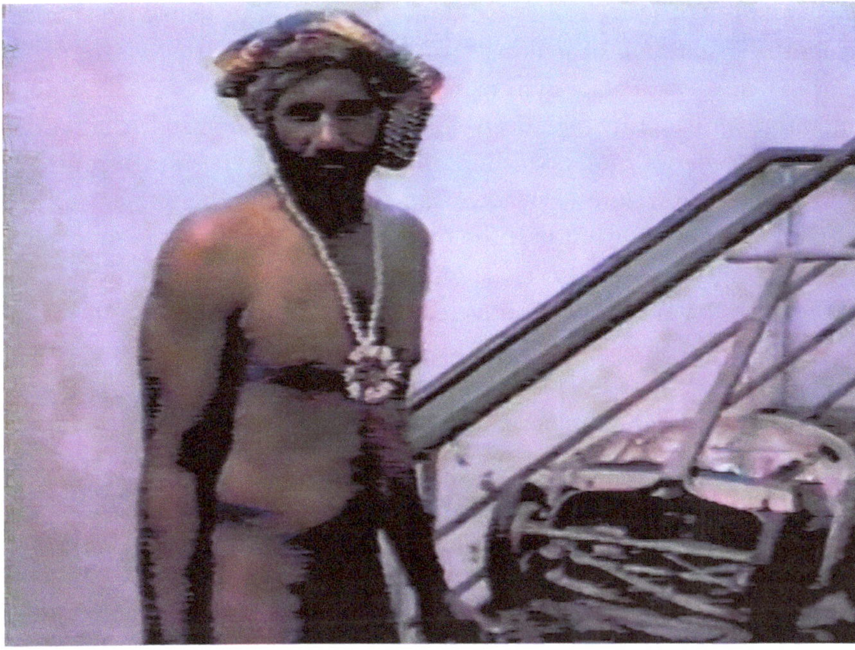

cut and provide the ground for new growth. Referring to himself as a griot, a West African storyteller, poet, and musician, a keeper of oral history and tradition, Jenkins announces a future reckoning with truth that will effect real estate, "investments and stockholders," and "psychological equity." The city referred to in the work's title is depicted as a heterogeneous network of styles, spaces, activities, people, identities, and technologies, not blended in a multicultural melting pot but rather competing with and complementing each other in a freewheeling and cacophonous hallucination, a place both concrete and elusive.

Through such works, Jenkins effectively presents a new economy, a new set of relations of selfhood production. *Dream City* feels like an

3.20

Ulysses Jenkins, *Dream City*, video still, 1981. Courtesy Ulysses Jenkins.

appropriation of a vaudeville tradition, which, as Aria Dean writes in her insightful discussion of Jenkins's work, is a source of the "image problem" of Blackness, a place of minstrel and blackface where African American images and culture have been "manipulated and misused . . . in order to distort society's understanding of black life."[189] The image problem with regard to "blackness" is not, Dean explains, that the image "fails to correspond to reality, but that the image has partly crafted reality," that Blackness itself is an image. This problem operates in multiple directions, challenging the authority of both clear-cut identity categories and "black art" as a form of resistance. As Dean points out, "Jenkins and his Los Angeles contemporaries . . . were often accused of making art that was not political enough or 'black enough' due to their interest in new media and abstraction and their willingness to draw on sources from outside of the black tradition." For these artists, the task of a critical emancipatory artistic practice cannot be to replace a false or fraught representation with a new or "truer" one, thus perpetuating the objectification and commodification/fetishization of the Black subject, but rather to change the logic of subjecthood altogether, moving, as Jenkins did, toward "a non-ontological blackness." If, in a culture defined by racial oppression, "the black" is always already an impure other, an "impure product," it is the nature (or naturalization) of the thing that has to be transformed, the very process of reification that must be subverted.[190] In earlier works such as *Mass of Images* (1978; figure 3.21), *Two Zone Transfer* (1979), and *Just Another Rendering of the Same Old Problem* (1979), Jenkins's singular body becomes the locus of multiplicity, mutilation, and defiance, a narrator of his own subjection to mass mediation. "He has renounced his desire to just be himself," Dean explains. "Jenkins has renounced actual being for an acceptance of the *historical* being, the 'ontological totality.' He has already 'consent[ed] not to be a single being.'"[191] With *Dream* City, this "n/ontology" expands to the collective body of blackness and beyond. Even the collective fantastic subject will no longer be available for reification through codes of color. But, like the city, this body is neither "multicultural" nor "post-racial"—it is an impure "no-thing-ness," a personal and social being as evolving construct, as process. *Dream City* features a merry-go-round, a literal and symbolic allusion to the fair and the fairground, past and potentially future sites of a struggle over the autonomy and control of

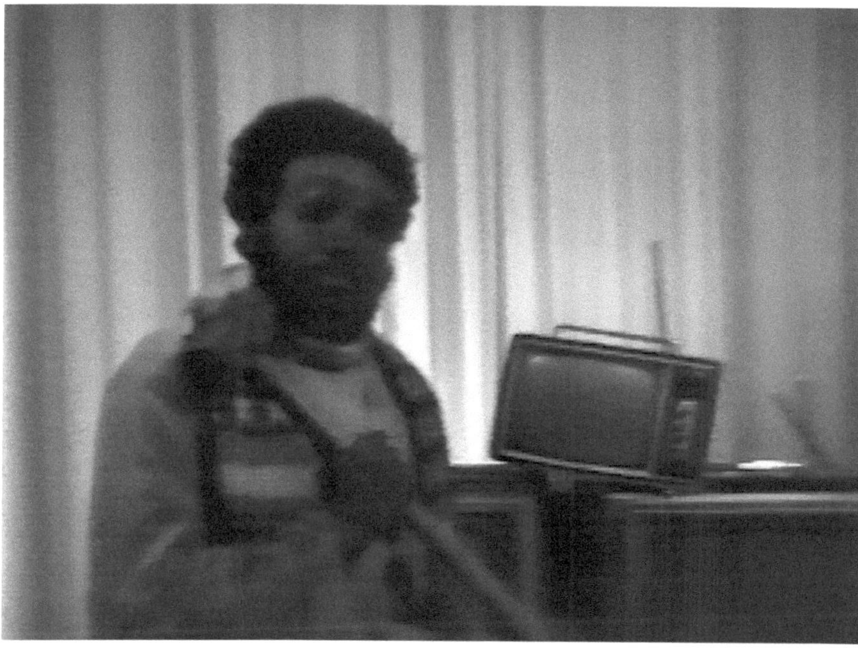

the body politic. As Jonathan Crary describes, whereas the modern carnivalesque retains some sense of destabilized identities, power structures, and divisions between spectator and performer, its "topsy-turvy world" has been divided off from the rationalized economic life of the city.[192] In *Dream City*, Jenkins refuses to segregate performing bodies from one another and from the electronic and physical space of the urban public and its population, while also refusing to bracket off art from the mediated and material experience of the everyday. The work does not provide a relegate-able resolution, a newly fixed embodiment of a fantastic constituency in space. The prophet remains ambivalent about the outlook: utopia

3.21

Ulysses Jenkins, *Mass of Images*, video still, 1978. Courtesy Ulysses Jenkins.

or dystopia depend on the process, the performance to come, not the (im)pure product of an existing technological and cultural logic.

In a similar sense, *Electronic Café* was less about establishing distinct sites for rational discourse than about creating a fairground-like experience, a carnivalesque place of interacting bodies and blurred boundaries, of often-marginalized peoples, of expansive performances and visionary materializations of past and future selves. This high-tech "café" was, in Stallybrass and White's terms, more tavern than coffeehouse, precisely the kind of place against which the bourgeois public sphere—with its antiseptic spaces for "refined" (read: disembodied) discussion—had been defined.[193] Or, more precisely, *Electronic Café* placed the coffeehouse in dialogue with the tavern. As mentioned, Günter's was explicitly modeled on the traditional coffeehouse, understood as "a forum for the arts and serious political discussion," while the three other restaurant locations were decidedly more "proletarian," popular places where the (racialized) body could not be denied or transcended: the "very rural, Korean-only" 8th Street Restaurant; the Gumbo House, decorated in an array of beer signs; the "all-Mexican" Ana Maria, with its old-country murals and waitresses in "peasant" dresses (figure 3.22). Adding to the fairground-like atmosphere, many *Electronic Café* participants staged live musical events, readings, poetry slams, and oral histories, bodies gathering on-site to give voice to and perform within and beyond given traditions, memories, and experiences. Others, in turn, uploaded images and narratives culled from private and public collections and archives in an attempt to document and speculate on how technological rendering functions as a device of both control and emancipation, sometimes presenting an image as a stifling confirmation of existing visual and ideological conventions, at other times as a platform for unexpected appearances, new presences, and affirmations of joyfully frenzied constructs of selves.

The network's central image database, the optical-disk extension of the original text-based Community Memory system, was an especially rich locus of such counterpublic endeavors, where intentional and sporadic contributions and uploads spawned a series of carnivalesque juxtapositions and composites. Examining this image archive, one finds an array of performances—some staged and rehearsed, others seemingly spontaneous—featuring costumes, masks, sets, encounters, overlapping

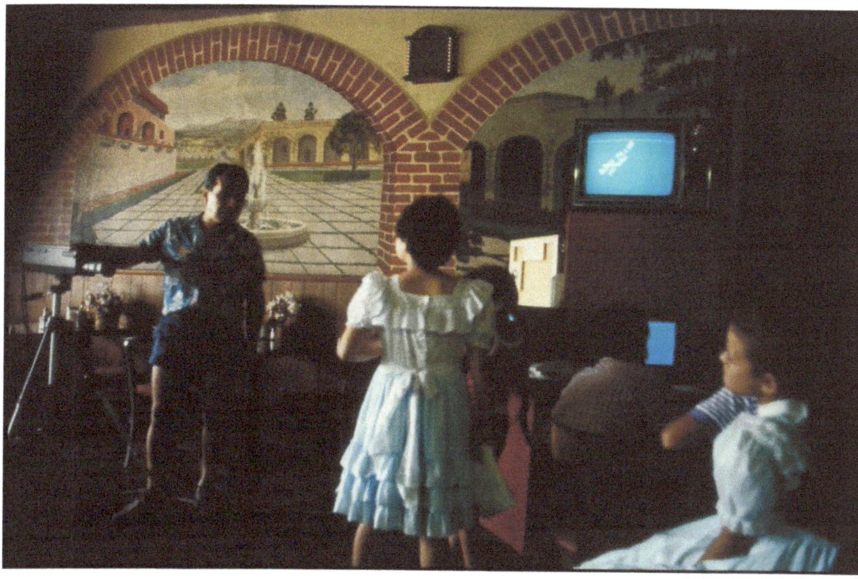

and overlaid fictional theatrics, and documentary materials. One of the most crucial aspects of this archive is that the difference between the purposefully constructed and the experienced is at times difficult to distinguish, causing an overall tension between the two, revealing their relationality as a dynamic, as a logic between real images and mediated reality. Rather than simply used as a platform for what would be considered proper public content and conduct, *Electronic Café* served as a site for the performance of transformation and transgressions, at times indistinguishably blending place and spaces, the physical and the virtual, the found and the made. One sequence of stills shows a young Latino man transforming himself into a clown, sitting in front of a mirror, carefully applying makeup, and putting on a wig and a hat. At some point, the screen

3.22

Mobile Image, *Electronic Café*, Ana Maria site (East LA), 1984. Courtesy Rabinowitz/ Galloway Archive.

splits into four asynchronous parts, blurring the moment between before and after, between reality, image, mask, dress, and role, while the viewer and performer encounter one another in the mirror, alluding to the instruments of seeing oneself and seeing someone else. The clown then leaves the room to walk into Ana Maria, mingling and posing with the guests who, in turn, look at the camera and at the viewer in front of the painted and sculpted walls of a restaurant designed to provide an intimacy and familiarity through a presentness of tradition. Preceding this chain of images are scanned cover art and announcements for the techno-punk band Nervous Gender and photographs of small children, a car with a smashed front, and a person in a t-shirt printed with a bull's-eye. Another sequence juxtaposes superimposed faces and features typically used to inscribe and ascribe places of origin and belonging, images of casual, joyful dancing recorded on-site in one of the restaurants, and close-up soft-core images of male genitalia.

Produced on-site and retrievable at the various nodes, such imagery would have further contrasted with the live television feeds that were a part of each console and which were dominated by the two big news stories of the moment: the 1984 presidential campaign and the Olympic Games—the latter also being the sponsor of *Electronic Café*. Both the President and the Games appear in the archive; one notable example features a screenshot of Reagan scratching his head. On top of this image, someone scrawled "Health, Housing, Education. YES. NO," a snap poll intended for others in the network. Yet, even when users did not explicitly incorporate these topics, they were continuously present, both as general context and as television broadcast. The montage of performances, bodies, texts, and images that was *Electronic Café* clashed with the controlled multiculturalism of the Olympics, with their composed, choreographed, uniformed, classical-humanist bodies that speak only to publicly accepted notions of physical health, social care, and competition on a level playing field. Such notions broadly aligned with the ideals of the Reagan era, with its neoliberal relegation of public health problems such as the AIDS crisis to a conveniently segregable sphere and idea of the "private" (private bodies, private sex, private problems) absent from public view and concern. Other sequences preserved in the *Electronic Café* archive include a performance at the Gumbo House featuring what looks like the reenactment of a ritual dance complete with traditional outfits, a display that functions as a Black counternarrative to the humanist-Eurocentric history embodied by the Olympic spectacle, as well as a critical reflection on the conventions of seeing that identify particular bodies, clothes, and movements as those of particular identities. This dynamic is underscored by stills showing the Black audience members at the Gumbo House, who appear to turn their gazes from the camera to the bodies in front of the console, working the instruments that both dematerialize and inscribe presence and experience. Similarly, at the 8th Street Restaurant, a masked figure is seen holding various objects up to the camera: first a Korean book on shaman ritual ceremony, then a skull, then an image of a nuclear bomb. A final shot of someone directly facing the skull as if deep in conversation slowly gives way to an image of the baldheaded Nosferatu, the 1979 version of the vampire played by Klaus Kinski. This sequence constitutes another estranging, poetic rumination on the oppressive and liberating powers of the dynamics of the material and immaterial dimensions of being as historical construct. The several celebratory transgressions of gender stereotypes likewise affirm the overall counterpublic force of the *Electronic Café* project, as historical figures in anachronistically extravagant dress are mixed with brightly colored faces that sport exaggerated lips, cheeks, and eyes and other features that in their exaltation no longer conform to assigned markers of difference. The normative is revealed as a matter of historical specificity and relativity, and the trans-body functions as a progressive politics of the subject, of any subject and subjects and the choice to conform or refuse to be confined to the spaces, forms, and expectations ascribed to them.

As the one explicitly institutional location in the network, the Museum of Contemporary Art, Los Angeles played a distinct role in Mobile Image's attempt to model a form of counterpublicity rather than solely expand on the existing spaces and dynamics of the bourgeois public sphere. The museum served as a somewhat ironic affirmation of the institutionalization of the struggle over the general horizon of social experience. Functioning as both a node and as a designated space for the accumulation, exhibition, and contemplation of the materials produced and exchanged in the network, it was there that those materials were most prominently collected, rematerialized, and displayed. Along with one of the consoles, the central image database was located in a mezzanine area off the lobby, and visitors to the museum were invited to peruse, add to, work on, and print out the contributions gathered on the server and tack them to the wall. Although these activities were present at all of the project's locations, MOCA was the only site inherently pre-inscribed as supra-local, hence designated as public, and thus seemingly subject to different rules of engagement than its distinctly neighborhood-based nodes. Photographs show numerous people who ostensibly came to see the museum's presentations of publicly relevant cultural artifacts working at the terminal among vast numbers of printouts. Those who would have ordinarily been designated (and self-identified) as "visitors" became producers. The wall collages of impressions and expressions, memories and imaginaries, emphatically juxtaposed the various self-assumed, socially ascribed, and culturally specificity publics with one another and with an official public institution of supposed collective relevance. They also foregrounded the constructedness of subjectivity itself. Users from the other sites also visited the museum in person to experience and partake in the arrangement and rearrangement of materials there. The results ranged from simple horizontal series and playfully chaotic clusters to coherent forms. One accumulation combined the aforementioned image of Reagan with the poll superimposed over his face with documentation of a visiting African dignitary at the Gumbo House and a General Motors advertisement that read "Nobody sweats the details like GM." Another contained a magazine cover featuring the story of an East LA "assault with a deadly weapon" with a series of Mexican vintage postcards and an *Electronic Café*–produced headshot of a man with another poll, asking: "Is this man . . . A. An inhalation therapist B. An East LA homeboy C. A UCLA PhD candidate in linguistics D. All the above E. None of the above" (figure 3.23). Another was a composite recreation of one of the other, restaurant-based consoles, stitched together from more than twenty eight-by-ten printouts. Most vividly, two enormous humanoid constellations were constructed from individual body-part images produced in the network. These figures, resisting normative markers of race, gender, and sexuality, towered over the space: in one arrangement, two bodies clad in fetish fashion were mounted with their backs facing the viewer, either walking away or pressed up against the wall (figure 3.24);

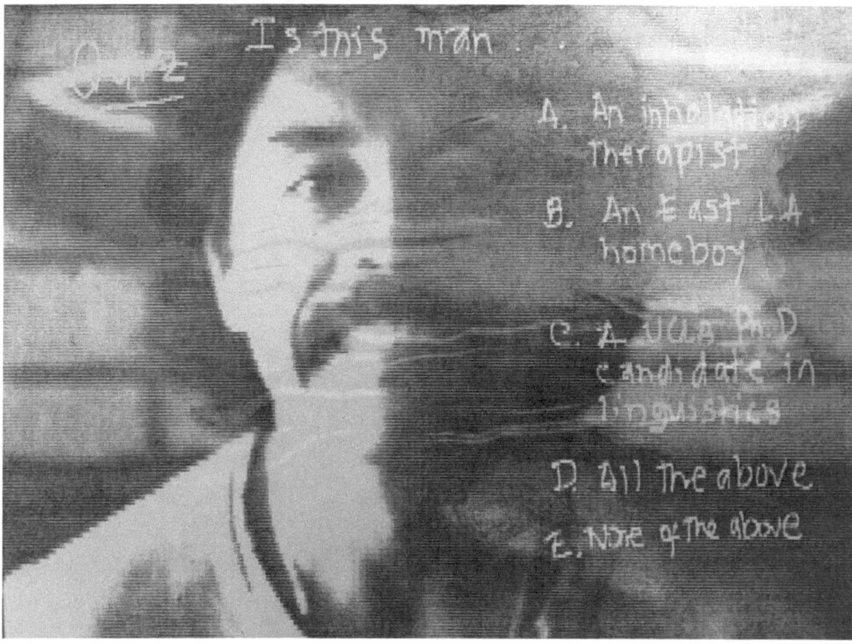

in the other a composite figure was stretched out to absurd proportions, arms held out wide in a gesture of embrace, flight, and proud self-display (figure 3.25). The counterpublic body was thus materialized: cyborgs composed of but not limited to their parts, each engaging with and defined by the ones next to it and the overall constellation without being subsumed by a preordained totality or meaning. Such a being is counterpublic in its historically specific process as it forms sporadically, strategically, politically according to particular present and future needs, not a product outside or substitute for itself. Thus, not surprisingly, the museum was unwilling or unable to maintain or contain this body and other assemblages similarly produced on-site. Without consulting Mobile Image, and

3.23

Mobile Image, *Electronic Café*, printout, 1984. Courtesy Rabinowitz/Galloway Archive.

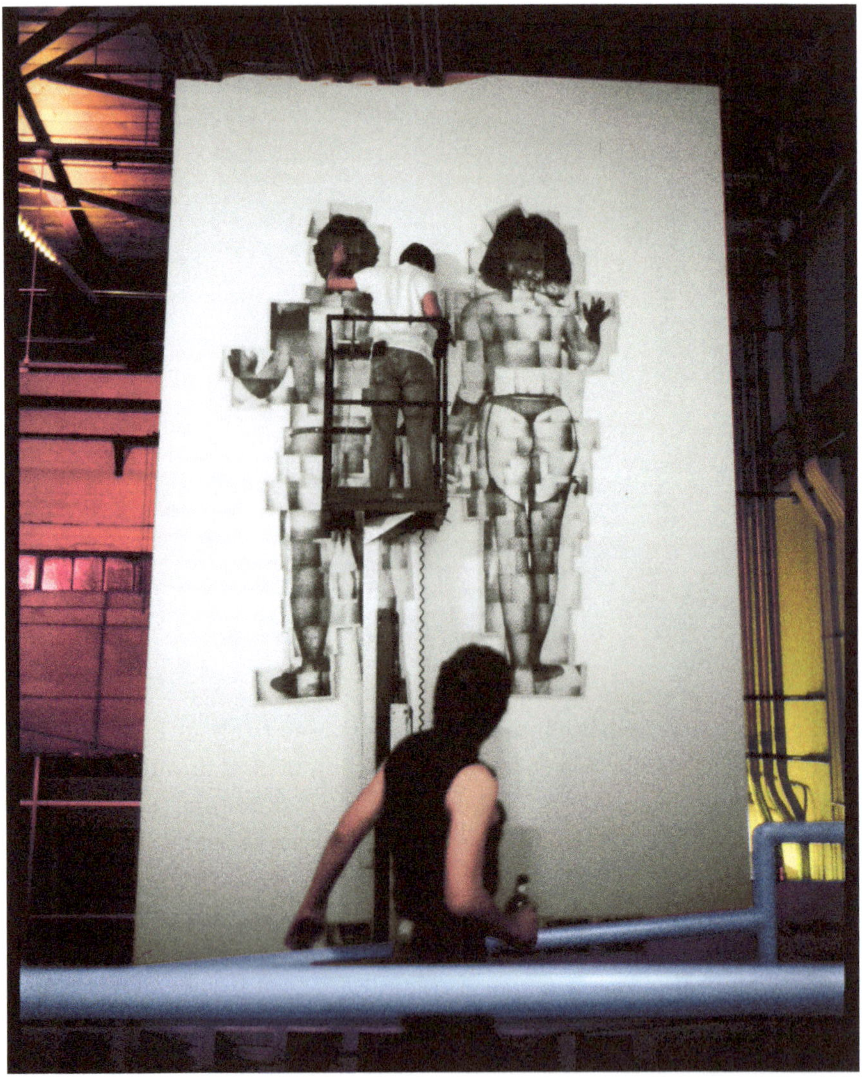

3.24

Mobile Image, *Electronic Café*, Museum of Contemporary Art location, 1984. Courtesy Rabinowitz/Galloway Archive.

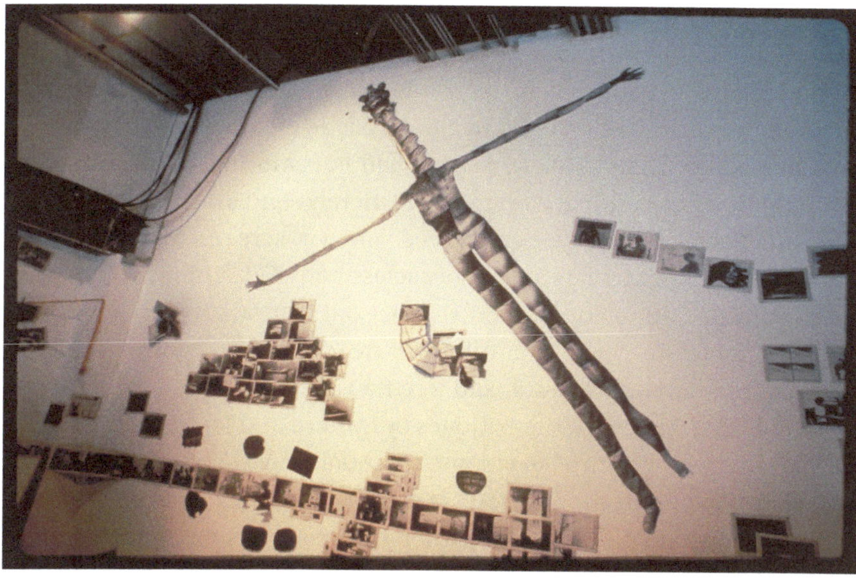

much to the artists' consternation, MOCA simply tore them down and dumped them in the trash at the end of the show's run.

ELECTRONIC COUNTERPUBLIC

Electronic Café engaged, at least in part, the same cultural and communicative crisis articulated in contemporaneous theories of postmodernism, particularly those that focused on an increasingly technologized world. Along with Jameson, late 1970s and early 1980s writers such as Jean Baudrillard and Jean-Francois Lyotard characterized post-1960s society as one dominated by computerization, electronic information, media

3.25

Mobile Image, *Electronic Café*, Museum of Contemporary Art location, 1984. Courtesy Rabinowitz/Galloway Archive.

spectacle, and multinational capitalism, resulting in a dissolution of subjectivity and authentic expression and the disappearance of a productive public sphere. To Baudrillard, the pandemic spread of advanced communication networks had ushered in a narcissistic era, in which consumers exist in isolation. The domestic space had become "a living satellite," a thoroughly privatized space where users merely send and receive signals. Life thus becomes empty, experienced only via screens and terminals, and "the real itself appears as a large useless body."[194] Whereas alienation once motivated opposition and resistance, all is now subsumed, in Baudrillard's mournful view, by an illusory freedom via a "pornography of information," abundant, fluid, and free-flowing.

Somewhat less bleak, the critiques by Lyotard and Jameson offer similar assessments but with at least some potential for certain kinds of agency. To Lyotard, "computerized society" is marked by a disintegration of grand narratives, with no possibility for either universal consensus or the establishment of a "pure" alternative to the current system. This "postmodern condition" emerges alongside an increasing concentration of power in the hands of those who control information and thus the economic system and knowledge production itself. In response, Lyotard calls for a "quite simple" solution to the problem of computerization: free public access to memory and data banks, which would enable a multiplicity of open discussions and self-conscious moments of consensus informed by inexhaustible reserves of information.[195] Jameson speaks specifically to the "aesthetic dilemma" facing the postmodern artist: in a technologized society drained of authenticity and subjectivity, it is no longer clear what the function of art is.[196] Mass culture responds with pastiche, epitomized by the science-fiction movie *Star Wars* (1977), with its appropriation of bygone forms in nostalgic revival of a prior generation's future imaginings. The only viable option for "high art," according to Jameson, is for practitioners to speak through "dead styles," rendering art a rumination on itself.[197] Indeed, so much art of the period was understood in those very terms.[198]

Mobile Image concurred with some of these core diagnoses—the saturation of society by prepackaged broadcast content, the excessive privatization of communication, the persistence of myths of technological progress and illusions of freedom, and the resulting threat to an operational public sphere. But they rejected the nihilism epitomized by

Baudrillard's theory, even as they openly rebutted industry-peddled promises of surefire social progress. To them, telecommunication networks were neither inherently democratizing nor part of a totalizing spectacle of control that leaves the user helpless to intervene.

Yet, crucially, *Electronic Café* represented an approach that was also distinct from the constructive options presented by theorists such as Jameson and Lyotard. Along with his call for increased access, the latter saw works of art as symbolic of budding sensibilities and thus capable of eliciting philosophical reflection on a computerized age. In 1985, Lyotard cocurated "Les Immatériaux," a landmark multimedia exhibition at the Centre Pompidou meant to highlight the bewildering and destabilizing effects of emergent technologies—as he put it, the "incertitude about the identity of the human individual in his condition of such improbable immateriality."[199] Focusing on conventional mediums such as painting and film, Jameson abided by a similar conception of cultural production as symbol, whether a pastiche that encapsulates the postmodern or a reflexive re-presentation of the inauthenticity of artistic expression itself. For both thinkers, the artist is fundamentally a producer of images—material or immaterial—that embody current conditions and potentially provoke contemplation.

Electronic Café offered a very different mode of artistic production: a participatory and open-ended social labor that both raised consciousness about the politics of technology and modeled ways to use it to effect social transformation. Participants glimpsed a counter-technological order, in which they could not only transmit their personal expressions and localized concerns but also partake in complex relations of struggle and negotiation. Alongside personal stories and calls for peaceful coexistence, exchanges about electoral politics, activism, religion, pornography, sexism, and other topics proliferated on the Community Memory bulletin board. "The person that entered the 'Nobody for President' obviously is very cynical and without hope," read one post. "I would like to enter into dialogue with him/her." "Why is it that no one admits they are a sexist?" asked another, to which someone responded, "Because they are not sure of their gender." At one point, a participant transmitted pictures and a narrative of himself having been beaten by police, triggering a discussion about police brutality against Black people in Los Angeles. Even as new bonds were forming, additional conflicts were emerging, such as the aforementioned exchange between the poets. These interactions enabled a vision not of a communalist utopia of frictionless communication but rather of critical and productive fantasy as a subversive, mobilizing process for challenging present ideological and material arrangements.

Hence, *Electronic Café* belongs not only to the history of techno-activism but also to the legacy of post-Minimalist aesthetics, particularly conceptual art, which was at least partially defined by an increased emphasis on what Alexander Alberro describes as "the possibilities of publicness and distribution."[200] When artists and critics of the late 1960s sought the dematerialization of the art object, it was not solely to overcome the restrictive commodification imposed by aestheticized, physical form. It was also to enable and demand participation, an opening up of art that, as Lucy Lippard and John Chandler put it in 1967, offers a "curious kind of Utopianism" whose "tabula rasa" would provoke a thinking through of ideas determined by the artwork as a catalyst or device wielded by those experiencing it.[201] According to Sol LeWitt, the idea "becomes the machine that makes the art," conjuring a technological metaphor that posits the viewer as an integral, ideally active part of the imaginative process.[202] And because, in this conceptualist schema, ideas are always shared and meaning created in that process, aesthetic production is ostensibly depersonalized—and therefore accessible, participatory, and public, upending hierarchical notions of skill and authorship in favor of communicative action. According to Benjamin Buchloh, however, this apparent democratization of means ultimately turned into a de facto "structural relationship of absolute equivalents," in which the quantity of information exchanged supplanted the potentially critical quality of aesthetic experience. Consequently, Buchloh argues, conceptualist works often replicated rather than transgressed the late-capitalist "logic of administration," thus reinforcing the institutions in which art serves as a mechanism for merely symbolic liberation, in which "artistic production is transformed into a tool of ideological control and cultural legitimation."[203]

Electronic Café can be understood as having heeded but not succumbed to Buchloh's critique. The work relates to several artistic practices of the 1970s that built on conceptualism but challenged some of its core tenets. In particular, *Electronic Café* resembles what Alberro calls an "antithetical model," exemplified by post-conceptualist artists such as Martha Rosler, Fred Lonidier, Allan Sekula, and Phel Steinmetz, who, with their activist, photo- and text-heavy practices, set out to show that "self-determination and communication, even in advanced forms of capitalist control, is still a historical option and artistic possibility."[204] What differentiates their work from that of other more pessimistic peers is that it critiques systems of representation for their ideological foundations while also reasserting the possibility to intervene politically within existing institutions. It refuses to cede the emancipatory potential of communication in the contemporary public sphere.[205] *Electronic Café* likewise problematized the presumed neutrality of various technologies of representation—linguistic, visual, technical, and otherwise—in order to elicit a conscious counterpublic refunctioning of productive and distributive tools.

Raising crucial questions about public life in the digital age, *Electronic Café* encouraged users to envision alternate arrangements of media control and enabled transgressive forms of public exchange. As Negt and Kluge explain, dominant publicity actively marginalizes nondominant groups by excluding them from its legitimizing power, which determines what can be said and how and whose experiences are considered relevant.[206] A successful counterpublicity needs to work productively beyond the reproductive ideological mechanisms of inclusion–exclusion. Set against the conformism of technological consumption and the myth of inherently rebellious forms of collaboration through creative heterogeneity, *Electronic Café* brought users together specifically so that they could recognize the ways in which normative participation in the "freedoms" offered by current telecommunication practices actually facilitated their own marginalization. Mobile Image modeled a form of counterpublicity designed to upend this arrangement and the power sustained by it. As Gene Youngblood describes, the effects were profound:

> The meetings became community events at which a great deal of serious discussion occurred among people who had never contemplated these possibilities before, and the identity of *Electronic Cafe* gradually emerged. Initial strategies included topics of discussion, solicitation of solutions to common problems, exchanges of cultural icons and symbols, translation of wit and wisdom from one language to another, photo dramas, collaborative pictorial creations, and various

games intrinsic to the visual and simultaneous drawing/writing components of the network. This participatory approach to the idea of "human design" is as much the point of *Electronic Cafe* as the network itself: for the first time in the history of electronic telecommunications, the identity of a large-scale, state of the art network issues from the vernacular language of indigenous culture, not the commodity jargon of corporate capitalism—an environment created and controlled by those who populate it.[207]

As a form of counterpublicity, *Electronic Café* facilitated instances of solidarity and reciprocity grounded in collective experiences of marginalization and exclusion. Yet, due to its particular setup—its mix of futuristic design and available technologies, commercial broadcast chatter and ad hoc creativity, uncharted electronic networks and familiar locales, prescribed cultural differences with experiences of common social concerns and myths of universal humanism with unexpected, historically specific needs—these instances were also highly reflexive. Participants were made aware of the mediated nature of their exchanges, of the fact that their communication was no longer rooted in traditional forms of social interaction and face-to-face relations, and that such exchanges were therefore subject to heightened evaluation, conflict, and negotiation.[208]

Galloway and Rabinowitz looked to the history of non- or pre-broadcast radio for related examples of critical and productive fantasy. They saw early amateur radio, for instance, as an archetype of politically effective, "utilitarian" communication technology. As Dieter Daniels explains, in the 1920s, radio was understood as having an "open, communicative, and networked structure," by which hobbyists could meet and exchange ideas over the air.[209] Here was a case of an emerging technology functioning as a highly localized tool serving the needs and future imaginings of particular publics. Whereas this function was soon superseded by centralized, mass distributed broadcast media, it survived into the later twentieth century within certain niche communities and served as a model for *Electronic Café*. As Galloway explains:

Cab drivers used radio, ham operators with their weird culture. They save people's lives, they do stuff. It's not like an artist doing things between here and there and sending something. These people are actually building the technology, launching their own mini satellites, and saving drowning people. That's utilitarian. We wanted to bring that word into the role of the artist in the technological society also. You could do things that actually help people. So we were creating these models and saying try it, you'll like it. It's not here yet, and you can't perpetuate it. You can't replicate *Electronic Café* installations with a $12,000 printer in each location. It's coming, get ready. Own this with your imagination so when the guy comes knocking on your door selling you a future, you kind of know what you want, rather than buying it. It was education.[210]

Mobile Image's approach to space, place, and communication also resembled the more explicitly political "micro-radio" movement of the 1980s. As theorized by artist-activist Tetsuo Kogawa, micro-radio combined the technology's de-territorializing capabilities with a decidedly territorial conception of the public sphere, involving an arrangement of highly localized communication nodes—a revival of Brecht's famous call for a "vast network of pipes." Through workshops, written treatises, and instructional videos and manuals, Kogawa championed "narrowcasting" via inexpensive parts and weak airwaves initially meant for devices such as remote-control toys and wireless microphones. "Micro" here connotes not just small size, low power, and limited range—usually only a 100- to 500-meter radius—but also a "micro politics," in Felix Guattari's sense of the term, Kogawa explains, comprising a qualitatively different function of radio technology, one that "comes back to the authentic meaning of 'techne' the old Greek of art, that is hand-work." Kogawa imagined a critical and productive "polymorphous" micro-radio network made up of overlapping transmissions publicizing the varied needs and concerns—both common and divergent—of local communities made up of multifarious subjects. As he puts it, "Today, the notion of individual is too large, to say nothing of 'group' and 'people'. Inside the individual, many singularities are buzzing and expressing themselves. In order to respond to such singularities, we need micro and diverse medium."[211]

Both *Electronic Café* and Kogawa's micro-radio work had more in common with earlier radio-based political interventions than with seemingly like-minded ventures of the time—the largely privatized exchanges of "telematic" art, as well as other works of radio art that, as discussed in chapter 1, sought to materialize the mass media apparatus or use it as a platform for distributed authorship or a source of readymade sound. *The Voice of Fighting Algeria* began in the mid-1950s, offering an alternative to the broadcasts of the French authorities and thereby giving rise to a shared anticolonial experience of solidarity. The radio, according to Frantz Fanon, begat the very possibility of a shared perspective, a fantasy of collective participation in liberation: "Having a radio meant paying one's taxes to the nation, buying the right of entry into the struggle of an assembled people."[212] A revolutionary consciousness was built through what was being said and heard over the ether and through the exercise of the public sphere as a site of active and engaged contention. As if to underscore the ongoing battle, the audience had to work their way along the dial to find the station, whose operators were continuously trying to evade the authorities' attempts to jam their frequency. Two decades later, the Italian "free radio" movement consisted of local stations occupying the public airwaves as a mechanism of subversive politics.[213] Bologna's Radio Alice was one of the first and most overtly activist of these efforts, transmitting an anti-institutional, Dadaist mix of diverse content, including poetry, lectures, readings, music, talk, prank phone calls, live coverage of protests, and cooking recipes.[214] Envisioned as "an artistic object" in and of itself, the station sought to create a community of otherwise marginalized groups—student, feminist, queer, worker—by approaching radio not as a homogenizing force but rather, in the words of cofounder Franco Berardi, as a "point of intersection of different experiences." In addition to its eclectic content, Radio Alice encouraged listeners to call in to the station and participate in the broadcast, opening radio to "the possibility of not only the circulation of information but also the circulation of struggle," as Berardi explains.[215] Although it was relatively short-lived—it went on air in 1976 and was closed by authorities in 1977—the station inspired both Guattari's call for "millions and millions of potential Alices"[216] and Kogawa's plan for a polymorphous radio network soon after.[217]

It is in relation to this context, and in connection to the utilitarian history of the radio referred to by Galloway, that the political edge of *Electronic Café* comes further into focus. And the ideas and notes compiled by Mobile Image and cited at the outset of this chapter have to be read as a dialectical inquiry into the politics of communications technology, informing and informed by, applied and performed by the work. Akin to Guattari and Kogawa's micro-politics, *Electronic Café* consisted of the charged connections among publics—marginal, partial, dominant—in a process of self-conscious positioning, as reach and context, local and trans-local, drawing and redrawing the lines that describe the circumference of community, public, interest, common ground, and shared experience became a technical as well as sociopolitical charge. As part of this process, the artists' notes regarding "human scale–technological scale," "realities and virtual realities" assert and reveal themselves as activated negotiations rather than binaries. When Galloway and Rabinowitz inquire about "the democratic process," they do not seek more or different technical devices and know-how in order to afford greater access to the existing democratic process, its public sphere and its apparatuses; rather, technology becomes a tool for the exploration of the democratic process as such, as participation and agency in the struggle over production of situations and positions, knowledge and meanings, experiences and policies. To use Haraway's words, "Technologies are skilled practices. How to see? Where to see from? What limits to vision? What to see for? Whom to see with? Who gets to have more than one point of view? Who gets blinded? Who wears blinders? Who interprets the visual field? What other sensory powers do we wish to cultivate other than vision?"[218] The reflective performance of these questions as a creative process politicizes the stakes set by the artists at the outset of their project: "Transform individual intelligence and social effectiveness? . . . The real issue is that of allowing access to the whole picture of what is going on."

AFTERMATH

As the culmination of more than a decade of Mobile Image work, *Electronic Café* was above all an act of relating. It did not offer a blueprint for a better tomorrow kept at arm's length, the proverbial carrot in front of the cart, but instead a self-conscious, politicized fantasy via concrete, real-time acts of counterpublicity. Its implications were absorbed by many who experienced it, and, in the years since, artists and activists from the communities involved produced an array of like-minded and far-reaching projects. Alhough varied in their objectives and techniques, these projects have critically engaged emergent commercial communication and networking technologies as sites of struggle, creating functional, participatory platforms for experiences of relationality. Heeded by some and overlooked by many, the lessons of *Electronic Café* became ever more urgent in a cultural context increasingly defined by the mass popularization of those technologies, fueled by entrenched myths of progress, expectations of progressive upheaval and transformation, and euphoric claims about an emergent electronic public sphere. Today, as platforms such as Facebook and X (formerly Twitter) dominate social and political interaction, we reckon with the consequences of hyper-privatized public discourse. The drawing pads, tablets, and scanners of *Electronic Café* may now be obsolete, but the work's implications remain profoundly relevant.

Electronic Café had immediate and lasting impact on Ben Caldwell, the engineer-in-residence at the Gumbo House in South LA and a member of LA Rebellion, a loose collective of filmmakers that formed in the early 1970s. LA Rebellion combined avant-garde cinema techniques with imagery and narratives representing the pressing concerns and politics of marginalized communities.[219] By the mid-1980s, Caldwell had become interested in Brecht's ideas on communication and the political efficacy of technologies, and *Electronic Café* reinforced those ideas by modeling what Caldwell calls a "pedestrian" approach: "It doesn't have the helicopters and the police and everything around that makes it seem like this omnipotent thing. It's just a pencil, you know, that's all it is. And it showed people, it's just a pencil and you can use it." For Caldwell, *Electronic Café* opened up possibilities for productive communication and enabled him to "see in a network sense":

> It blew me away. I thought "wow, this is the answer," because that's what I was interested in in filmmaking, is how can we emancipate our image, . . . the image of who we were . . . It was really another way to kind of look at busting down doors, so you could go to another neighborhood that you couldn't normally access and communicate directly, person to person, with them over this medium and exchange ideas and concepts. So I started working that way . . . The walk-in capacity of [*Electronic Café*] really made me see that we could be like doctors in the neighborhood, and we could take in clients right straight off the street and answer their needs. Whatever that need is.[220]

Later in 1984, Caldwell founded KAOS Network, a still-operational multimedia community production, training, and exhibition center located on a prominent corner in Leimert Park, a hub of Black art and culture in Los Angeles. The activities hosted at the center aim to demystify preexisting technologies and educate local residents on how to use them in alignment with their own objectives, connect with others inside and outside their communities, and think about the relationship of such technologies to seemingly all-powerful institutions and systems. In 1986, Caldwell inaugurated "I-Fresh Express," a multi-year KAOS-based platform, involving projects designed to "pass on mass communications skills to black youth."[221] In the decades since, KAOS has hosted numerous programs, from hip-hop open-mic sessions to multimedia exchanges between youths in Los Angeles and Cuba and South Africa. The recent *Leimert Phone Company* (2013) repurposed obsolete public pay phones to construct a neighborhood-wide network representing the voices and needs of the community (figure 3.26).[222] Although some of these projects have retained notions of democratization via technological know-how and access, others have more strategically redeployed existing technologies pointedly to engage the very processes of urban development and exclusion, the ongoing struggles over the participation in and production of relations, experiences, and policies.

Caldwell remained connected to Galloway and Rabinowitz, who themselves continued to build on the lessons of their previous work. In 1988, they founded Electronic Café International (ECI), a "permanent public telecommunications lab where we could connect with other people and build an international network," as Rabinowitz put it.[223] Housed at the 18th Street Arts Center in Santa Monica, California, ECI took advantage of the increasing availability of networking technologies; it was, for Mobile Image, the next logical step in "the creation of a virtual social commons where the convergence of intimate terminal space, public space, and virtual space comes together as an experience." As they explain, "After the opening of the original *Electronic Cafe Network* in 1984 we felt that we had reached 'the limits of models.' All of our previous work begged to be developed. The next step was 'community,' a permanent multimedia collaborative public network. In 1988 ECI opened with its first international link with Paris.—ECI creates a networked lab, to support collaboration and

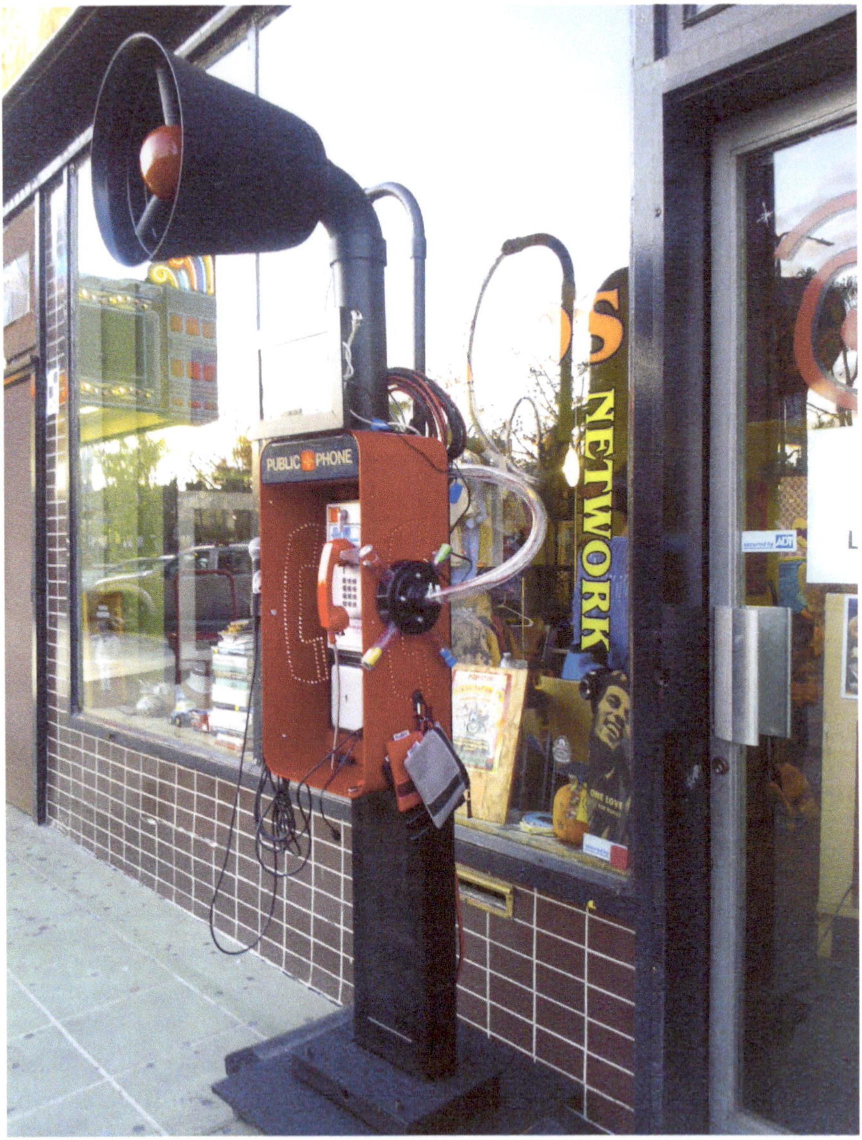

3.26

Leimert Phone Company, "Sankofa Red" prototype outside KAOS Network, Leimert Park, Los Angeles, 2013. Courtesy Ben Caldwell.

co-creation between people in different cultures, countries and language groups."[224] As with the 1984 *Electronic Café*, visitors could experiment with emergent technologies and form new publics with people near and far, while the permanence of ECI allowed them to integrate those technologies into their lives more fully. Users could set up their first personal email account or schedule regular "virtual hangouts." They could also form communities around like-minded interests and goals. In 1989, a group of regulars created Tele-Poetry, a videophone-based poetry network connecting ECI to venues in New York City, Boston, and eventually other locations in the United States and abroad. Artists and musicians also produced a range of multimedia "telecollaborative" performances and concerts via the network. By 1991, ECI had more than thirty networked affiliates in countries around the world, including Brazil, Denmark, Israel, Canada, France, and Japan.

Most poignantly, a series of explicitly political ECI activities emerged, projects that reminded users of broader contexts and their own positions within the network. Such activities revealed how certain tools can be wielded not only to create new artforms and construct new solidarities, new publics, but also to reimagine given technological arrangements of bodies, knowledge, place, and experience. In 1990, Ana Coria, an original *Electronic Café* community-outreach organizer, traveled to Nicaragua and set up a videophone connection between LA-based Nicaraguan-Americans and journalists and prominent Sandinista revolutionaries and author Omar Cabezas in an effort to both confront and subvert mass-mediated perceptions of South American politics and the people associated with them (figure 3.27).[225] In 1991, Ulysses Jenkins produced *Bay Window: A Videophone Ritual Performance*, which linked together native Canadian communities in Western Front, Vancouver, and Baker Lake, Hudson Bay, with the Headlands Center for the Arts in Sausalito, California, ECI's Santa Monica location, and the Exploratorium in San Francisco (figure 3.28). The event was designed to connect disparate indigenous communities to each other and to other marginalized groups, raise awareness about mutual concerns, and combine diverse modes of knowledge production and transference. It reorganized relationships between the supposedly traditional and the high tech and between various supposedly distinct "ethnic" identities, forms, and issues. The succession

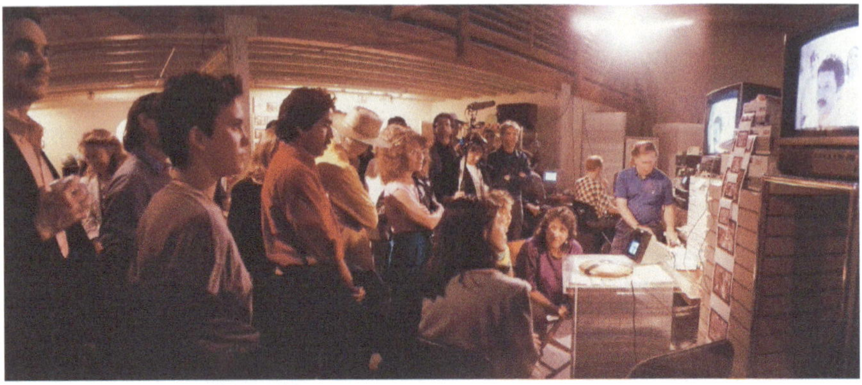

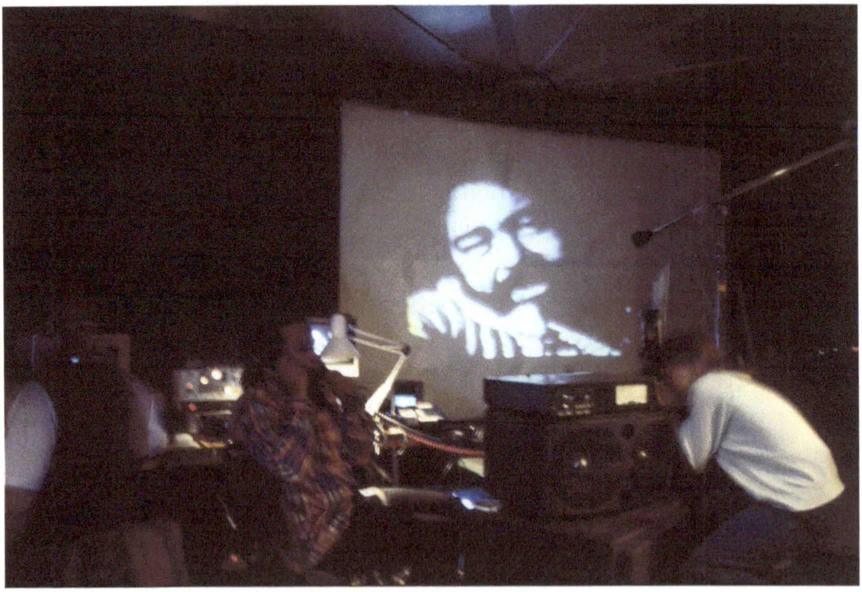

3.27

Mobile Image, Electronic Café International, event with Omar Cabezas, Nicaragua, 1990. Courtesy Rabinowitz/Galloway Archive.

3.28

Ulysses Jenkins, *Bay Window: A Videophone Ritual Performance*, Jenkins (foreground, left) in San Francisco, conversing with a Native Canadian man (on screen) in Baker Lake, Canada, 1991. Courtesy Ulysses Jenkins.

of performances, storytelling, images, and discussions included footage from a live Miwok harvest ceremony, a political statement by a member of Southern California's Native American community, Spanish-language folk songs, a performance by the Collage Ensemble, a Los Angeles–based group of Japanese-American sound artists, the creation of a sand painting, a tribute to Anwar Sadat, a free-jazz performance, and a conversation on the planned construction of an open-pit uranium mine in Nunavut territory and the expected environmental impacts. Interspersed among these happenings were technical and logistical comments by operators of the network, feedback, visual and aural glitches, and the sound of modem signals, effectively materializing the mechanical apparatus. The following year, Caldwell and Jenkins produced *Video Phone to South Africa*, a three-way audio and video conference that connected local youths in Santa Monica and Oakland, California, with members of the ANC in South Africa. In 1994, ECI hosted "Café Barbie," an examination the doll's cultural significance by participants in Paris, New York, and Santa Monica.[226]

In their conception of art as laboratory for experimentation, education, and experiences of relationality, endeavors such as these and those at KAOS and other venues can, like the earlier work of Mobile Image, be understood in terms established by the historical avant-garde. Tools not typically used for artistic practice were similarly reconceived as such, with the goal of both enabling new connectivities or new symbols and converting those tools into functional means of social, political, and perceptual innovation. Aesthetic transformation would thus beget a new technics of aesthetics based on the needs of diverse publics continuously being formed and re-formed via technological networks. As Brecht explained, such an approach has the potential to alter the landscape of communication and knowledge production and the role of individuals within such networks radically, to "bring [them] into a relationship instead of isolating [them] . . . turning the audience not only into pupils but into teachers." Post–*Electronic Café* works such as those discussed above facilitated such experiences of relationality by redistributing communication tools according to temporary alliances formed around common concerns and needs, including the experience of exclusion and the mechanisms that separate the officially public from the presumably private.

This mode of artistic production distinguishes this legacy from what, at first glance, might seem to be like-minded participatory, technologically oriented art ventures that emerged around the same time. In 1986, for example, Carl Loeffler and Fred Tusk established ACEN (Art Comm Electronic Network), a dial-up BBS conceived as an "electronic gathering place" for the creation and dissemination of art projects and information.[227] Such endeavors led to what Dieter Daniels terms "frameworks," a series of electronic artist networks meant to serve as permanent infrastructures for open, text-based communication. Formed in the early 1990s, when personal computers and modems were becoming increasingly available but before the Internet became a true mass medium, these projects included *Public Netbase* in Vienna (1994), *Internationale Stadt Berlin* (1994) and, most notably, *THE THING* (1991), an international BBS network founded by Wolfgang Staehle in New York, with independent nodes soon after established throughout Western Europe. Conceived as high-tech versions of Josef Beuys's notion of "social sculpture," the frameworks attempted to realize, as Staehle put it, "Beuys's idea of direct democracy, of a political community as a social structure."[228] These projects represented a utopian ideal of what Ries terms "pure sociality," reflecting a euphoria over the possibility of limitless, autonomous communication."[229]

The frameworks paralleled what is commonly termed relational aesthetics, a group of contemporaneous, primarily "offline" practices that nonetheless adopted various networking paradigms of the moment—participation, communication, collectivity, sociality. In his theorization of the movement, Nicolas Bourriaud not only borrowed technological terminology such as "user friendliness" and "interactivity,"[230] but also understood relational art as signifying "the changing mental space that has been opened for thought by the Internet, the central tool of the information age we have entered."[231] Although he was determined to distinguish this movement from the art of the 1960s and 1970s, Bourriaud's assertions recalled core claims made on behalf of conceptualism, emphasizing dematerialization, expanded participation, and a leveling of hierarchies.[232] Echoing Sol LeWitt, he considered the work of relational art "a machine for provoking and managing individual or collective encounters."[233] And whereas Bourriaud was convinced that utopian thinking was obsolete, he saw such artworks as "micro-utopias," moments of frictionless communalism, of open democratic exchange, built upon perceived commonalities.[234]

As critics have pointed out, however, this conception of art—like that of both earlier conceptualist practices and the concurrent framework networks—generally conforms to the logic of dominant systems, of a globalized economy built upon expanded communication.[235] Gene Ray puts a finer political point on such critiques, arguing that Bourriaud's micro-utopianism effectively enacts a "de-radicalization" of historical avant-garde politics, an abandonment of the "macro" aspiration to transform systems of social relations and overcome exploitation and domination.[236]

The most politically potent network-based works enable users to break free of those limitations by not only exchanging personal expressions and local concerns but also reaching beyond conventional institutions, situating themselves within broader struggles, and actively participating in those struggles. What distinguishes *Electronic Café* and works like it is a critical practice that is at once analytic, reflexive, and transformative, squarely aimed at effecting systemic change. This practice is itself an argument about, as Ray puts it, "the whole legacy—and so also the present and future—of the avant-garde project." It has also become increasingly vital in the "Internet Age," and especially with the rise of social media, by which individuals everywhere can broadcast their views, becoming so-called prosumers. Society is only now beginning to reckon with a technological apparatus dominated by a small group of corporations and characterized by widespread misinformation, distraction, privacy breaches, and election meddling. And yet, the potential to redeploy mass media tools on behalf of the marginalized or dispossessed remains, and certain artists and activists continue to use those tools to offer critical and productive fantasy as a subversive, mobilizing process for challenging current ideological and material arrangements, a generator of truly innovative communication.

Our point here is not to elevate Mobile Image above all others or advance simplistic notions of influence or credit. As discussed, *Electronic Café* was itself a collaboration, and some of its artists- and engineers-in-residence had already been involved with, and played a role in developing, the approaches epitomized by that landmark project. One of our primary objectives has been to sketch an art history that adds nuance to extant accounts of "art and technology"—including histories of new media and digital art and much broader understandings of how art and politics relate to the tools by which they are brought into being. In such an art history, "communication" and "technology" are neither seen as inherently democratizing nor organized into symbolic displays of connectivity, of fabled transcendence, detached from material conditions and the historical specificities of the user and his or her context, the politics of time and territory. In this sense, we aim to follow the very methodology modeled by Mobile Image, entering into the historical record, the "community memory," certain overlooked projects, perspectives, and narratives—art history as itself an act of counterpublicity. To varying degrees, the activities that succeeded *Electronic Café* were conceived as laboratories for the strategic repurposing of existing technologies on behalf of specific social needs—of fantasy, as Negt and Kluge understood it—grounded in historically rooted "situated knowledges" and the relationality of the public and the private. They attempted to generate new forms of knowledge and practice, new networks of bodies, information, place, and agency, via the *Umfunktionierung* of available tools.

Certainly, there are other artists and projects that we have not acknowledged, some with no connection to Mobile Image and yet no less worthy of study. What is most important is to underscore a particular approach, one that mobilizes fantasy to facilitate experiences of counterpublicity. In the contest over what and who are considered relevant, in the struggle over ownership, access, and authority and the quest for social transformation, the politicization of technology has the potential to redefine critical collective engagement. The works of Mobile Image involved self-conscious positioning and repositioning, as definitions of place, community, public, and interest were recast as technical as well as sociopolitical processes. Participants encountered not just novel devices but also substantive shifts in relationality in the ways in which communication and experience are organized through elaborate arrangements of tools and their use. Although largely overlooked by extant histories of art, these works and those like it have become ever more relevant. They provide crucial contributions to contemporary discourse, nuancing artistic categories such as "new media" and "social practice," while also engaging broader political and ideological struggles facing a world of instant digital interaction, social media, and consolidated corporate control.

NOTES

INTRODUCTION

1. Mobile Image, "Concepts and Ideas," unpublished notes, n.d., n.p., the Sherrie Rabinowitz and Kit Galloway Archive (hereafter referred to as the Rabinowitz/Galloway Archive).
2. Walter Benjamin, "The Work of Art in the Age in the Age of Its Technological Reproducibility" (1936–1939), in *Walter Benjamin. Selected Writings, 4: 1938–1940*, ed. Howard Eiland and Michael W. Jennings (Cambridge, MA: Harvard University Press, 2003), 255.
3. Gene Youngblood, "Metadesign: Towards a Postmodernism of Reconstruction" (1986), quoted in Dieter Daniels, "Reverse Engineering Modernism with the Last Avant-Garde," in *Net Pioneers 1.0: Contextualizing Early Net-Based Art*, ed. Dieter Daniels and Gunther Reisinger (Berlin: Sternberg Press, 2009), 15.
4. The discussion of avant-garde strategies and practices and their ongoing relevance could be found in some of the leading publications at the time. See the discussion between T. J. Clark and Peter Wollen in *Screen* (T. J. Clark, "Preliminaries to a Possible Treatment of Olympia in 1865," *Screen* 21, no. 1 [Spring 1980]: 18–51; Peter Wollen, "Manet: Modernism and Avant-Garde. Timothy Clark's Article on Manet's *Olympia, Screen*, Spring 1980," *Screen* 21, no. 2 [Summer 1980]: 15–25); Rosalind Krauss, "The Originality of the Avant-Garde: A Postmodernist Repetition" *October* 18 (Fall 1981): 47–66; the translation and publication of Peter Bürger's *Theory of the Avant-Garde* (Minneapolis: University of Minnesota Press, 1984); and Benjamin Buchloh's critical review "Theorizing the Avant-Garde," *Art in America* 72, no. 10(November 1984): 19–21.
5. Gene Ray, "Avant-Gardes as Anti-Capitalist Vector," *Third Text* 21, no. 3 (May 2007): 241.
6. Gene Youngblood, *Expanded Cinema—Fiftieth Anniversary Edition* (New York: Fordham University Press, 2020), xxv.
7. See Gene Youngblood, "Virtual Space: The Electronic Environments of Mobile Image," *International Synergy Journal* 1, no. 1 (1986): 9–20; and "Virtual Space: The Electronic Environments of Mobile Image," in *Ars Electronica: Facing the Future (A Survey of Two Decades)*, ed. Timorty Druckrey (Cambridge, MA: MIT Press, 1987), 360–365. We have had access to some of these unpublished sections—co-written with Galloway and Rabinowitz—from the Rabinowitz/Galloway Archive. We reference those drafts with as much specificity as possible in our text.

8. Rhizome, "Net Art Anthology," accessed October 17, 2022, https://anthology.rhizome.org/mobile-image.

9. Rhizome, "Net Art Anthology."

10. Annemarie Chandler, "Animating the Social: Mobile Image/Kit Galloway and Sherrie Rabinowitz," in *At a Distance: Precursors to Art and Activism on the Internet*, ed. Annemarie Chandler and Norie Neumark (Cambridge, MA: MIT Press, 2005), 163.

11. Chandler, "Animating the Social," 167.

12. Chandler, "Animating the Social," 172.

13. "The Art of Participation: 1950 to Now," San Francisco Museum of Modern Art, November 8, 2008–February 9, 2009.

14. Rudolf Freiling, "Introduction," in *The Art of Participation: 1950 to Now*, exhibition catalog, San Francisco Museum of Modern Art (New York: Thames & Hudson, 2009), 12–14.

15. Tanya Zimbardo, catalog entry, *The Art of Participation: 1950 to Now*, 140.

16. Tom McDonough, "The Art of Participation: 1950 to Now," *Artforum International* 47, no. 8 (April 2009): 180–181.

17. "Collaboration Labs: Southern California Artists and the Artist Space Movement," 18th Street Art Center, October 1, 2011–March 16, 2012, part of "Pacific Standard Time: Art in LA 1945–1980."

18. Alex Donis, "California Dreamin': Performance, Media Art, and History as Gossip," in *Collaboration Labs: Southern California Artists and the Artist Space Movement*, exhibition catalog (Santa Monica, CA: 18th Street Art Center, 2011), 19, 67–69.

19. Julia Bryan-Wilson, "Alternative Media Landscapes in Los Angeles in the 1970s and 1980s," in *Collaboration Labs: Southern California Artists and the Artist Space Movement*, exhibition catalog (Santa Monica, CA: 18th Street Art Center, 2011), 71–73.

20. Bryan-Wilson, "Alternative Media Landscapes," 72–73.

21. Bryan-Wilson, "Alternative Media Landscapes," 83.

22. Kris Paulsen, *Here/There: Telepresence, Touch, and Art at the Interface* (Cambridge, MA: MIT Press, 2017), 10–11.

23. Paulsen, *Here/There*, 10–13, 115–118.

24. Paulsen, *Here/There*, 10–13, 119.

25. Bertolt Brecht, "The Radio as an Apparatus of Communication," in *Brecht on Theater*, ed. John Willett (New York: Hill and Wang, 1996), 52.

26. Brecht, "The Radio," 53.

27. Mobile Image, "Three Experiments: Exploring the Aesthetics of Satellite Communications" (1977), unpublished manuscript, 4 (Rabinowitz/Galloway Archive).

28. Brecht, "Radio—An Antediluvian Invention?" (1927), in *Bertolt Brecht on Film and Radio*, trans. and ed. Marc Silberman (London: Methuen, 2000), 36–38.

29. Mobile Image, "Three Experiments," 4.

30. Lazlo Moholy-Nagy, "Production—Reproduction" (1922), reprinted in Krisztina Passuth, *Moholy-Nagy* (London: Thames & Hudson, 1985), 289.

31. Moholy-Nagy, "Production—Reproduction," 289.

32. Moholy-Nagy, "Production—Reproduction," 289.

33. Christina Kiaer, "Boris Arvatov's Socialist Objects," *October* 81 (Summer 1997): 105–118.

34. Steven Durland, "Defining the Image as Place: A Conversation with Kit Galloway, Sherrie Rabinowitz, and Gene Youngblood," *High Performance* 37 (1987): n.p.

35. Mobile Image, "Concepts and Ideas" (emphasis/all caps in original).

36. Galloway in Durland, "Defining the Image as Place" (authors' emphasis).

37. Mobile Image, "Concepts and Ideas."

38. Throughout the 1970s and 1980s, writers such as Peter Wollen, Griselda Pollock, Sylvia Harvey, and Stuart Hall critiqued the logic that had governed the reception and debates concerning the historical avant-garde and its legacy, arguing that even advocates of the so-called neo-avant-gardes followed the historical trajectory of a politics of aesthetics that disregarded radical ideological and socioeconomic transformation through public agency. See Peter Wollen, "The Two-Avant-Gardes," *Studio International* 190, no. 978 (November/December 1975): 171–175; Griselda Pollock, "Screening the Seventies: Sexuality and Representation in Feminist Practice—A Brechtian Perspective," in *Vision and Difference; Femininity, Feminism, and Histories of Art* (London: Routledge, 1988), 155–199; Sylvia Harvey, "Whose Brecht? Memories for the Eighties," *Screen* 1 (1982): 45–59; Stuart Hall, "Notes on Deconstructing 'The Popular,'" in *People's History and Socialist Theory*, ed. Raphael Samuel (London: Routledge, 1981), 227–240.

39. Youngblood in Durland, "Defining the Image as Place" (authors' emphasis).

40. Peter Thompson, "Introduction," in *The Privatization of Hope: Ernst Bloch and the Future of Utopia*, ed. Thompson and Slavoj Žižek (Durham, NC: Duke University Press, 2013), 5.

41. Patrick D. Hazard, ed., *TV as Art: Some Essays in Criticism* (Champaign, IL: National Council of Teachers of English, 1966); Douglas Davis and Allison Simmons, eds., *The New Television: A Public/Private Art* (Cambridge, MA: MIT Press, 1977); Ivan Illich, *Tools for Conviviality* (New York: Harper and Row, 1973); Raymond Williams, *Television: Technology and Cultural Form* (first published in 1974) (London: Routledge, 1990).

42. Herbert Schiller, *Communication and Cultural Domination* (White Plains, NY: International Arts and Sciences Press, 1976), 7.

43. Gene Youngblood, *Expanded Cinema* (New York: Dutton, 1970); Hans Magnus Enzensberger, "Constituents of a Theory of the Media" (1970), trans. Stuart Hood, in *Critical Essays*, eds. Reinhold Grimm and Bruce Armstrong (New York: Continuum, 1982), 46–76; Oskar Negt, "Mass Media: Tools of Domination or Instruments of Liberation? Aspects of the Frankfurt Schools' Communications Analysis" (1973), trans. Leslie Adelson, *New German Critique* 14 (Spring 1978): 61–80.

44. Enzensberger, "Constituents of a Theory of the Media," 50.

45. Enzensberger, "Constituents of a Theory of the Media," 56.

46. Ernst Bloch, *Experimentum Mundi: Frage, Kategorien des Herausbringens, Praxis* (Frankfurt: Suhrkamp, 1975), 11.

47. Ernst Bloch, *The Principle of Hope*, vol. 1 (Cambridge, MA: MIT Press, 1986), 195.

48. Bloch, *The Principle of Hope*, 196.

49. Bloch, *The Principle of Hope*, 201.

50. Bloch, *The Principle of Hope*, 196.

51. Mobile Image, "Concepts and Ideas."

52. Oskar Negt and Alexander Kluge, *The Public Sphere and Experience: Toward an Analysis of the Bourgeois and Proletarian Public Sphere* (Minneapolis: University of Minnesota Press, 1993), 1–2 (emphasis in original).

53. Peter Hohendahl, "Jürgen Habermas: 'The Public Sphere' (1964)," *New German Critique* 3 (Autumn 1974): 47. Habermas's study, *The Structural Transformation of the Public Sphere: An Inquiry into a Category of Bourgeois Society*, was originally published in German in 1962. It first appeared in translation in the United States in its entirety in 1991 (Cambridge, MA: MIT Press). Excerpts and shortened versions were available to an American audience at least as early as 1974, when the *New German Critique* published Habermas's "The Public Sphere: An Encyclopedia Article (1964)" (Autumn 1974): 49–55.

54. Negt and Kluge, *Public Sphere and Experience*.

55. Miriam Hansen, "Foreword," in *Public Sphere and Experience*, xxx.

56. Galloway and Rabinowitz were aware of the potential reproduction of the public sphere's dichotomous logic: "Interactive networks will be a tremendous social organizing force—whether such networks will tend to integrate or further fragment social structure is an open question, but an important issue" (Mobile Image, "Concepts and Ideas").

57. Mobile Image, "Concepts and Ideas."

58. Mobile Image, "Concepts and Ideas" (emphasis/all caps in original).

59. Oskar Negt and Alexander Kluge, *Geschichte und Eigensinn* (1981) (Frankfurt: Suhrkamp, 1993), 371 fn. 16. See also Fredric Jameson, "On Negt and Kluge," *October* 46 (Fall 1988): 156, fn. 5.

60. Negt and Kluge, *Public Sphere and Experience*, 18.

61. Negt and Kluge, *Public Sphere and Experience*, 3.

62. For a discussion of the new social movements, see Stanley Aronowitz, "The New Social Movements and Class," in *How Class Works: Power and Social Movement* (New Haven, CT: Yale University Press, 2003), 141–170.

63. Homi Bhabha, "How Newness Enters the World: Postmodern Space, Postcolonial Times and the Trials of Cultural Translation," *The Location of Culture* (London and New York: Routledge, 1994), 219 (emphasis in original).

64. bell hooks, *Feminist Theory: From Margin to Center* (Boston: South End Press, 1984), xvi.

65. Daniel Martinez, interview with the authors, Los Angeles, March 2, 2016.

66. Donna Haraway, "Situated Knowledges: The Science Question in Feminism and the Privilege of Partial Perspective," *Feminist Studies* 14, no. 3 (Autumn 1988): 584.

67. Haraway, "Situated Knowledges," 584.

68. José Esteban Muñoz, *Cruising Utopia: The Then and There of Queer Futurity* (New York: New York University Press, 2009), 1.

69. Muñoz, *Cruising Utopia*, 28, 26.

70. Bloch, cited in Muñoz, *Cruising Utopia*, 27.

71. David Cateforis, Steven Duval, and Shepherd Steiner, *Hybrid Practices: Art in Collaboration with Science and Technology in the Long 1960s* (Oakland: University of California Press, 2019), 1.

72. Cateforis et al., *Hybrid Practices*, 2.

73. John Beck and Ryan Bishop, *Technocrats of the Imagination: Art, Technology, and the Military-Industrial Avant-Garde* (Durham, NC: Duke University Press, 2020), 1.

74. Beck and Bishop, *Technocrats*, 1.

75. Beck and Bishop, *Technocrats*, 1–2.

76. Donna Conwell and Glenn Philips, "Duration Piece: Rethinking Sculpture in Los Angeles," in *Pacific Standard Time: Los Angeles Art 1945–1980*, exhibition catalog, eds. Rebecca Peabody, Andrew Perchuck, Glenn Phillips, and Rami Singh (Los Angeles: Getty Research Institute and the J. Paul Getty Museum, 2011), 209.

77. John Beck and Ryan Bishop, "The Return of the Art and Technology Lab," *Cultural Politics* 14, no. 2 (2018): 226.

78. Maurice Tuchman, *Introduction to a Report on the Art and Technology Program of the Los Angeles County Museum of Art 1967–1971* (Los Angeles: Los Angeles County Museum of Art, 1971), 9.

79. Beck and Bishop, *Technocrats*, 127.

80. Beck and Bishop, *Technocrats*, 8, 127–128.

81. Tuchman, *Introduction*, 17.

82. Cateforis et al., *Hybrid Practices*, 3.

83. Anne Collins Goodyear, "Launching 'Hybrid Practices' in the 1960s: On the Perils and Promise of Art and Technology," in *Hybrid Practices: Art in Collaboration with Science and Technology in the Long 1960s* (Oakland: University of California Press, 2019), 37.

84. Goodyear, "Launching 'Hybrid Practices' in the 1960s," 37.

85. Beck and Bishop, *Technocrats*, 168.

86. Beck and Bishop, *Technocrats*, 8–12.

87. Beck and Bishop, *Technocrats*, 108.

88. W. Patrick McCray, "Fallout and Spinoff: Commercializing the Art-Technology Nexus," in *Hybrid Practices: Art in Collaboration with Science and Technology in the Long 1960s* (Oakland: University of California Press, 2019), 62.

89. Erin Manning and Brian Massumi, *Thought in the Act: Passages in the Ecology of Experience* (Minneapolis: University of Minnesota Press, 2014), 84.

90. Manning and Massumi, *Thought in the Act*, 85.

91. Manning and Massumi, *Thought in the Act*, 87.

92. Manning and Massumi, *Thought in the Act*, 88–89.

93. Manning and Massumi, *Thought in the Act*, 91–92.

94. Manning and Massumi, *Thought in the Act*, 111.

95. Rebecca Peabody et al., "Shifting the Standard: Reappraising Art in Los Angeles," in *Pacific Standard Time: Los Angeles Art 1945–1980*, exhibition catalog, ed. Rebecca Peabody, Andrew Perchuck, Glenn Phillips, and Rami Singh (Los Angeles: Getty Research Institute and the J. Paul Getty Museum, 2011), 3.

96. Julia Bryan-Wilson, "'Out to See Video': EZTV's Queer Microcinema in West Hollywood," *Grey Room* 56 (Summer 2014): 66.

97. Gene Youngblood, "Intro to Mobile Image Projects," unpublished text, n.d., 11 (Rabinowitz/Galloway Archive).

98. Beck and Bishop, "The Return of the Art and Technology Lab," 227–228.

99. Hito Steyerl, "Art as Occupation: Claims for an Autonomy of Life," *The Wretched of the Screen* (Berlin: Sternberg Press, 2012), 110.

100. Ray, "Avant-Gardes as Anti-Capitalist Vector," 241.

101. Beck and Bishop, "The Return of the Art and Technology Lab," 227.

102. Regarding the depoliticization of the Bauhaus in postwar art history, see Rainer Wick, "Rezeptionsgeschichtliche Randbemerkungen zur Bauhaus-Pädagogik," in *Ist die Bauhaus Pädagogik aktuell?* (Cologne: Walter König, 1985), 42–58; Philip Glahn, "The Bauhaus—Lost in Transfer: Art as Work," in *Wolkenkuckucksheim—International Journal of Architectural Theory* 24, no. 39 (2019): 151–162.

103. Melvin Kranzberg, "Technology and History: 'Kranzberg's Laws,'" *Technology and Culture* 27, no. 3 (July 1986): 545.

104. For a discussion of "project" versus "work-form," see Ray, "Avant-Gardes as Anti-Capitalist Vector," 247.

105. Harvey, "Whose Brecht?," 46.

106. Wick, "Rezeptionsgeschichtliche Randbemerkungen," 45 (authors' translation).

107. Wick, "Rezeptionsgeschichtliche Randbemerkungen," 47 (authors' translation).

108. Wick, "Rezeptionsgeschichtliche Randbemerkungen," 47 (authors' translation).

109. Walter Gropius, *Die Idee und Aufbau des Staatlichen Bauhauses Weimar* (Weimar, Germany: Bauhausverlag, 1923), 1 (authors' translation).

110. Gregory Sholette and Blake Stimson, *Collectivism after Modernism* (Minneapolis: University of Minnesota Press, 2007), 13.

111. Kit Galloway, Sherrie Rabinowitz, and Gene Youngblood, "Meta: Economic Support for Social Metadesign: The Avantpreneur," unpublished text, n.d., 3 (Galloway/Rabinowitz Archive).

112. Galloway, Rabinowitz, and Youngblood, "Meta: Economic Support for Social Metadesign."

113. Youngblood, "Intro to Mobile Image Projects," 1.

114. Rabinowitz, unpublished interview with Youngblood and Galloway, n.d., 6–7 (Rabinowitz/Galloway Archive).

115. *Screen*, the leading journal of film and television studies with significant impact on visual arts practice and discourse, was a central voice in the Brechtian reconsideration of a popular-activist culture and published two influential special issues on Brecht in the mid-1970s: "Special Number: Brecht and a Revolutionary Cinema" (Summer 1974) and "Brecht and the Cinema/Film and Politics" (Winter 1975–1976).

116. Harvey, "Whose Brecht?," 46.

117. Harvey, "Whose Brecht?," 46.

118. Harvey, "Whose Brecht?," 47.

119. Alain Badiou, *The Communist Hypothesis* (London: Verso, 2010); David Harvey, "The End of Capitalism?" Penn Humanities Forum Lecture, University of Pennsylvania, Philadelphia, PA, November 30, 2011.

120. WHW, "What Keeps Mankind Alive?" 11th Istanbul Biennial, Istanbul, Turkey, September 12–November 8, 2011.

121. Fredric Jameson, quoted in WHW, "What Keeps Mankind Alive?," 53.

122. WHW, curators' text, "What Keeps Mankind Alive?," 112.

123. "Mark Zuckerberg announces radical changes to Facebook," accessed June 22, 2020, https://newsbeezer.com/romaniaeng/mark-zuckerberg-announces-radical-changes-to-facebook-all-users-will-be-affected-news-by-source/.

124. Quoted in WHW, curators' text, "What Keeps Mankind Alive?," 102.

CHAPTER 1

1. Mobile Image, "Satellite Arts Project" (Rabinowitz/Galloway Archive).
2. "TV as a Creative Medium," May 17–June 14, 1969. For an overview, see Marita Sturken, "TV as a Creative Medium: Howard Wise and Video Art," *Afterimage* 11, no. 10 (1984): 5–9.
3. Mobile Image, unpublished notes (Rabinowitz/Galloway Archive).
4. Galloway and Rabinowitz, quoted in Youngblood, unpublished draft of manuscript, 27 (Rabinowitz/Galloway Archive).

5. Benjamin, "Work of Art," 255.
6. Miriam Hansen, "Unstable Mixtures, Dilated Spheres: Negt and Kluge's *The Public Sphere and Experience*, Twenty Years later," in *Public Culture* 5, no. 2 (Winter 1993): 210–211.
7. Youngblood, *Expanded Cinema*, 258; Williams, *Television: Technology and Cultural Form*, 149.
8. Enzensberger, "Constituents of a Theory of the Media," 46–47.
9. John G. Hanhardt, *The Worlds of Nam June Paik* (New York: Guggenheim Museum, 2000), 34–35.
10. For a contemporaneous overview of this activity, see Youngblood, *Expanded Cinema*, 257–316.
11. Youngblood, *Expanded Cinema*, 341–343.
12. Frank Gillette, quoted in Youngblood, *Expanded Cinema*, 343.
13. Michael Asher, *Writings 1973–1983 on Works 1969–1979*, written in collaboration with Benjamin H.D. Buchloh (Halifax and Los Angeles: Nova Scotia College of Art and Design and The Museum of Contemporary Art Los Angeles, 1983), 112–116.
14. Brice Howard, quoted in Youngblood, *Expanded Cinema*, 283.
15. Edward A. Shanken, *Art and Electronic Media* (London: Phaidon Press, 2009), 32.
16. Dierdre Boyle, "Guerilla Television Revisited," *Art Journal* 45, no. 3 (1985): 228–232.
17. Douglas Davis, interview, accessed January 13, 2023, http://afsnitp.dk/udefra/1/dd/museum.html.
18. Douglas Davis, "I just want to GO!," interview with Tilman Baumgärtel, Rhizome, accessed January 13, 2023, http://rhizome.org/community/41653/.
19. Paulsen, *Here/There*, 105.
20. Liza Béar and Keith Sonnier, "Send/Receive Phase I and II Documents: 1977," accessed January 13, 2023, http://sendreceivesatellitenetwork.blogspot.com/.
21. Paulsen, *Here/There*, 105–107.
22. Keith Sonnier, "Aesthesipol: Keith Sonnier by Betsy Sussler," interview in *BOMB*, Spring 1982, accessed January 28, 2016, http://bombmagazine.org/article/67/.
23. Enzensberger, "Constituents of a Theory of the Media," 49.
24. Allison Simmons, "Introduction. Television and Art: A Historical Primer for an Improbable Alliance," in *The New Television: A Public/Private Art* (Cambridge, MA: MIT Press, 1977), 7.
25. Williams, *Television: Technology and Cultural Form*, 139–140 (authors' emphasis).
26. Negt, "Mass Media," 64.
27. Marshall McLuhan, *Understanding Media: The Extensions of Man* (New York: McGraw-Hill, 1964), 265–274.
28. Stephen Kern, *The Culture of Time and Space: 1880–1918* (Cambridge, MA: Harvard University Press, 1983), 69.

29. McLuhan, *Understanding Media*, 269.

30. For a description of "Art by Telephone," see https://mcachicago.org/Exhibitions/1969/Art-By-Telephone.

31. See, for example, Friedrich Knilli, "Die Arbeiterbewegung und die Medien: Ein Rückblick," *Gewerkschaftliche Monatshefte* (06/1974): 349–362; Knut Hickethier, "Hitler und das Radio: Der Rundfunk in der NS-Zeit," in *Die Kultur der 30er und 40er Jahre*, ed. Werner Faulstich (Munich: Wilhelm Fink Verlag, 2009), 191–208.

32. Brecht, "Das Radio: Eine vorsinflutliche Erfindung?" (1927), in *Bertolt Brecht. Große kommentierte Berliner und Frankfurter Ausgabe*, ed. Werner Hecht et al. (Berlin: Aufbau and Suhrkamp, 1993), XXI, 217.

33. Brecht, "Radio as an Apparatus of Communication," 52 (authors' emphasis).

34. John Durham Peters, "Mass Media," in *Critical Terms for Media Studies*, ed. W. J. T. Mitchell and Mark B. N. Hansen (Chicago: University of Chicago Press, 2010), 273, 276.

35. Daniel Baird, "Processing Space," *The Brooklyn Rail*, November 1, 2003, accessed January 13, 2023, http://brooklynrail.org/2003/11/art/processing-space.

36. Max Neuhaus, "Public Supply I," *Media Art Net*, accessed January 13, 2023, http://rhizome.org/editorial/2010/aug/03/public-supply-i-1966-max-neuhaus/.

37. Max Neuhaus, "The Broadcast Works: Radio Net," accessed January 13, 2023, http://www.kunstradio.at/ZEITGLEICH/CATALOG/ENGLISH/neuhaus2b-e.html.

38. Dieter Daniels, "Re-Inventing Radio," in *Re-Inventing Radio: Aspects of Radio as Art*, ed. Heidi Grundmann et al. (Frankfurt am Main: Revolver, 2008), 44–45.

39. "World Ear Project by KPFA Radio," call for participation, *PAGE: Bulletin of the Computer Arts Society* 13 (January 1971).

40. "Close Radio," *High Performance* 1, no. 4 (December 1978): 12–15.

41. Mobile Image, "Notes on Performance Structure and Design" (Rabinowitz/Galloway Archive).

42. Mobile Image, unpublished notes (Rabinowitz/Galloway Archive).

43. John Durham Peters, *Speaking into the Air: A History of the Idea of Communication* (Chicago: University of Chicago Press, 1999), 5.

44. Stephan Oettermann, *The Panorama: History of a Mass Medium* (1980), trans. Deborah Lucas Schneider (New York: Zone Books, 1997), 44, 7.

45. Oettermann, *The Panorama*, 7.

46. Benjamin, "Work of Art," 255.

47. Jonathan Crary, "Géricault, the Panorama, and Sites of Reality in the Early Nineteenth Century," *Grey Room* 9 (Fall 2002): 21.

48. Crary, "Géricault," 22–23.

49. Caroline Jones, "Senses," in *Critical Terms for Media Studies*, ed. W. J. T. Mitchell and Mark B. N. Hansen (Chicago: University of Chicago Press, 2010), 97.

50. Schiller, *Communication and Cultural Domination*, 48.
51. Schiller, *Communication and Cultural Domination*, 25.
52. Schiller, *Communication and Cultural Domination*, 6.
53. Schiller, *Communication and Cultural Domination*, 3.
54. NASA, "A Public Service Communications Satellite Experiment," February 1977 (Rabinowitz/Galloway Archive).
55. Galloway, unpublished interview with Youngblood.
56. Paulsen, *Here/There*, 98–99.
57. Lisa Parks, *Cultures in Orbit* (Durham, NC: Duke University Press, 2005), 23.
58. Parks, *Cultures in Orbit*, 41.
59. Parks, *Cultures in Orbit*, 25–26 (authors' emphasis).
60. Donna Haraway, cited in Parks, *Cultures in Orbit*, 14.
61. Negt and Kluge, *Public Sphere and Experience*, 158.
62. Miriam Hansen, "Early Cinema, Late Cinema: Transformations of the Public Sphere," in *Viewing Positions: Ways of Seeing Film*, ed. Linda Williams (New Brunswick, NJ: Rutgers University Press, 1994), 143.
63. Alexander Kluge, "Why Should Television and Film Cooperate?" trans. Stuart Liebman, *October* 46 (Autumn 1988): 96.
64. Negt and Kluge, *Public Sphere and Experience*, 101.
65. Negt and Kluge, *Public Sphere and Experience*, 101–102.
66. Peters, "Mass Media," 274.
67. Judith Butler, *Notes Toward a Performative Theory of Assembly* (Cambridge, MA: Harvard University Press, 2015).
68. Rabinowitz, unpublished interview with Youngblood.
69. Paulsen, *Here/There*, 109.
70. Youngblood, unpublished draft (Rabinowitz/Galloway Archive).
71. Rabinowitz, unpublished interview with Youngblood.
72. Rabinowitz, "The Satellite Arts Project/some final notes and images" (Rabinowitz/Galloway Archive).
73. Richard Serra, *Television Delivers People* (1973), 7 min., single channel video.
74. Vilem Flusser, "Two Approaches to the Phenomenon Television," in *The New Television: A Public/Private Art* (Cambridge, MA: MIT Press, 1977), 237.
75. Flusser, "Two Approaches," 237.
76. Enzensberger, "Constituents of a Theory of the Media," 55.
77. Enzensberger, "Constituents of a Theory of the Media," 60–61.
78. Particia Mellencamp, "Prologue," in *Logics of Television: Essays in Cultural Criticism*, ed. Patricia Mellencamp (Bloomington: Indiana University Press, 1990), 3.
79. Hansen, "Early Cinema," 140.

80. Hansen, "Early Cinema," 142.

81. Bruce Clarke, "Communication," in *Critical Terms for Media Studies*, ed. W. J. T. Mitchell and Mark B. N. Hansen (Chicago: University of Chicago Press, 2010), 139–140.

82. Enzensberger, "Constituents of a Theory of the Media," 47.

83. Enzensberger, "Constituents of a Theory of the Media," 58–59.

84. Enzensberger, "Constituents of a Theory of the Media," 48.

85. Galloway and Rabinowitz, quoted in Youngblood, unpublished draft, 27 (Rabinowitz/Galloway Archive).

86. Galloway and Rabinowitz, quoted in Youngblood, unpublished draft, 27.

87. Tom Gunning, "The Cinema of Attractions: Early Film, Its Spectator and the Avant-Garde," in *Early Cinema: Space, Frame, Narrative*, ed. Thomas Elsaesser and Adam Barker (London: BFI, 1990), 59. Our argument is indebted to Miriam Hansen's significant expansion of Gunning's idea in "Early Cinema," 134–152.

88. Gunning, "Cinema of Attractions," 56.

89. Gunning, "Cinema of Attractions," 57.

90. Paulsen, *Here/There*, 112.

91. Mobile Image, "The Scores" (Rabinowitz/Galloway Archive).

92. Michael Shamberg and Raindance Corporation, *Guerilla Television* (New York: Holt, Reinhart and Winston, 1971), 9.

93. Shamberg, *Guerilla Television*, 12.

94. Deirdre Boyle, "From Portapak to Camcorder: A Brief History of Guerrilla Television," *Journal of Film and Video* 44, no. 1–2 (Spring–Summer 1992): 70.

95. Boyle, "From Portapak," 71.

96. Deirdre Boyle, "Subject to Change: Guerrilla Television Revisited," *Art Journal* 45, no. 3(Fall 1985): 232 (authors' emphasis).

97. Mobile Image, "Satellite Arts Project" (Rabinowitz/Galloway Archive).

98. Mobile Image, "The Scores."

99. Mobile Image, "The Satellite Arts Project/some final notes and images."

100. Mobile Image, "The Scores."

101. Mobile Image, "The Scores."

102. Mobile Image, "The Satellite Arts Project/some final notes and images" (authors' emphasis).

103. Bertolt Brecht, "Zur Theorie des Lehrstücks" (1937), in *Bertolt Brecht. Große kommentierte Berliner und Frankfurter Ausgabe*, ed. Werner Hecht et al. (Berlin: Aufbau and Suhrkamp, 1993), XXII, 351.

104. Brecht, "Zur Theorie," 351.

105. Paulsen, *Here/There*, 118.

106. Gunning, "Cinema of Attractions," 60.

107. Jacques Rancière, "The Emancipated Spectator," *Artforum* 45, no. 7 (March 2007): 279.

108. Youngblood, unpublished draft, 12.

109. Marita Sturken, "Paradox in the Evolution of an Art Form: Great Expectations and the Making of History," in *Illuminating Video: An Essential Guide to Video Art*, ed. Doug Hall and Sally Jo Fifer (New York: Aperture and Bay Area Video Coalition, 1990), 113.

110. Deirdre Boyle, "A Brief History of American Documentary Video," in *Illuminating Video: An Essential Guide to Video Art*, ed. Doug Hall and Sally Jo Fifer (New York: Aperture and Bay Area Video Coalition, 1990), 62.

111. Boyle, "A Brief History," 62.

112. Boyle, "A Brief History," 62–63.

113. For early, contemporaneous discussions of feminism and technology, see Shulamith Firestone, *The Dialectic of Sex: The Case for Feminist Revolution* (New York: Morrow, 1970); Laura Mulvey, "Visual Pleasure and Narrative Cinema," *Screen* 16, no. 3 (Autumn 1975): 6–18; Claire Johnstone and Paul Willemen, "Brecht in Britain: The Independent Political Film (on The Nightcleaners)," *Screen* 16, no. 4 (1975/6): 101–118; and Camera Obscura Film Collective, "Feminism and Film: Critical Approaches," *Camera Obscura* 1 (Fall 1976): 3–10.

114. Coco Fusco, "Ethnicity, Politics, and Poetics: Latinos and Media Art," in *Illuminating Video: An Essential Guide to Video Art*, ed. Doug Hall and Sally Jo Fifer (New York: Aperture and Bay Area Video Coalition, 1990), 305.

115. Cited in Fusco, "Ethnicity," 304.

116. Youngblood, unpublished draft, 12.

117. Alexander Galloway, *The Interface Effect* (Cambridge, MA: Polity Press, 2012), 17.

118. Galloway, *The Interface Effect*, 23.

119. Youngblood, unpublished draft, 12.

120. Lisa Nakamura, *Digitizing Race: Visual Cultures of the Internet* (Minneapolis: University of Minnesota Press, 2007), 30.

121. Wendy Hui Kyong Chun, "Digital and New Media," in *A Companion to Critical and Cultural Theory*, ed. Imre Szeman, Sarah Blacker, and Justin Sully (Hoboken, NJ: John Wiley, 2017), 396.

122. Paul Dourish, *Where the Action Is: The Foundations of Embodied Interaction* (Cambridge, MA: MIT Press, 2004), 18.

123. Nakamura, *Digitizing Race*, 2.

124. Nakamura, *Digitizing Race*, 3.

125. Mobile Image, "Concepts and Ideas."

126. For a recent report on "Diversity in High Tech," see https://www.eeoc.gov/special-report/diversity-high-tech (accessed April 21, 2022).

127. Lisa Parks, "Satellite and Cyber Visualities," cited in Nakamura, *Digitizing Race*, 15.

128. Nakamura, *Digitizing Race*, 16.

129. Nakamura, *Digitizing Race*, 20.

130. Paul Richard, "The Satellite Fantastic, or Dancing into Video History," *The Washington Post*, November 23, 1977, B4.

131. The first Black astronaut to travel to space as part of a NASA mission in 1983 was Guion Bluford. See Anna Leahy and Douglas R. Dechow, "What Everyone Gets Wrong About Black History in the Space Age," *Scientific American*, February 7, 2017.

132. Nakamura, *Digitizing Race*, 20.

133. Chela Sandoval, *Methodology of the Oppressed*, quoted in Nakamura, *Digitizing Race*, 20.

134. Nakamura, *Digitizing Race*, 35 (emphasis in original).

135. Hansen, "Early Cinema," 147.

136. Hansen, "Early Cinema," 147 (authors' emphasis).

137. https://www.melkweg.nl/en/info/about-melkweg/, accessed October 23, 2021.

138. Galloway, quoted in "Prelude to Satellite Arts," unpublished interview with Youngblood, 1–2.

139. Dara Greenwald, "The Grassroots Video Pioneers," *The Brooklyn Rail*, May 2, 2007, accessed January 13, 2023, http://brooklynrail.org/2007/05/express/video; Boyle, "Subject to Change," 229.

140. Bryan-Wilson, "'Out to See Video,'" 59, 61.

141. Bryan-Wilson, "'Out to See Video,'" 69.

142. Bryan-Wilson, "'Out to See Video,'" 59–60.

143. Michael Fried, "Art and Objecthood" (1967), reprinted in *Minimal Art: A Critical Anthology*, ed. Gregory Battcock (Berkeley: University of California Press, 1995), 129.

144. Rosalind Krauss, "Video: The Aesthetics of Narcissism," *October* 1 (Spring, 1976): 56.

145. For an earlier, brief discussion of *Satellite Arts* vis-à-vis Krauss's assessment of video's narcissist dimensions, see Paulsen, *Here/There*, 115–116.

146. Nathan Stinson, "Interview with Nathan" (Rabinowitz/Galloway Archive).

147. Mobile Image, "The Satellite Arts Project/some final notes and images."

148. Mobile Image, "The Satellite Arts Project/some final notes and images."

149. Stinson, "Interview with Nathan."

150. Stinson, letter to Kit Galloway and Sherrie Rabinowitz, February 13, 1978 (Rabinowitz/Galloway Archive).

151. Leslie A. Adelson, "Contemporary Critical Consciousness: Peter Sloterdijk, Oskar Negt/Alexander Kluge, and the 'New Subjectivity,'" *German Studies Review* 10, no. 1 (February 1987): 61.

152. Adelson, "Contemporary Critical Consciousness," 66.

153. Negt and Kluge, cited in Adelson, "Contemporary Critical Consciousness," 62.

154. Unpublished "evaluation/debriefing session" among participants (Rabinowitz/Galloway Archive).

155. Anna Halprin, "Three Decades of Transformative Dance. Interview by Nancy Stark Smith," in *Moving Toward Life: Five Decades of Transformational Dance* (London: University of Press of New England, 1995), 14.

156. Lawrence Halprin, *The RSVP Cycles: Creative Processes in the Human Environment* (New York: George Brazilier, 1969); Halprin, *Moving Toward Life*.

157. Mobile Image, notes on *Satellite Arts* (Rabinowitz/Galloway Archive).

158. Mobile Image, notes on *Satellite Arts*.

159. Mobile Image, "The Scores."

160. Unpublished "evaluation/debriefing session" among participants.

161. Mobile Image, "The Scores."

162. Butler, *Notes Toward a Performative Theory of Assembly*.

163. Christoph Menke, "Review: Judith Butler, *Notes Toward a Performative Theory of Assembly*," *Notre Dame Philosophical Reviews*, February 6, 2016, accessed January 20, 2023, http://ndpr.nd.edu/news/notes-toward-a-performative-theory-of-assembly/.

164. James Clifford, cited in Menke, "Review." For further discussion of theory as travel, translation, and reflexiveness, see Menke, "Philososphie der sinnlichen Kraft," *dctp.tv—Nachrichten Werkstatt*, October 18, 2016, http://www.dctp.tv/filme/philosophie-der-sinnlichen-kraft-newsstories-18102016/.

165. Butler, *Notes Toward a Performative Theory of Assembly*, 33.

CHAPTER 2

1. Quoted in Chandler, "Animating the Social," 164.
2. Galloway and Rabinowitz, "Hole-In-Space, 1980," accessed February 5, 2015, http://www.ecafe.com/getty/HIS/.
3. Mobile Image, artist statement (Rabinowitz/Galloway Archive).
4. Tim Wu, "The Tyranny of Convenience," *New York Times*, February 18, 2018, accessed February 18, 2018, https://www.nytimes.com/2018/02/16/opinion/sunday/tyranny-convenience.html.
5. Galloway, unpublished interview with Youngblood.
6. Mobile Image, unpublished interview with Youngblood.
7. Rabinowitz, unpublished interview with Youngblood. The videographers were Bill and Este Marpet in New York and Lynn Adler and Jules Backus in Los Angeles.
8. Mobile Image, *Hole in Space* artist statement (Rabinowitz/Galloway Archive).

9. Rabinowitz, unpublished interview with Youngblood.
10. Youngblood, unpublished draft.
11. Pamela Lifton-Zoline, "Notes for 'Hole-in-Space: A Communication Sculpture,'" unpublished essay, 1980 (Rabinowitz/Galloway Archive).
12. Mobile Image, unpublished interview with Youngblood.
13. Lifton-Zoline, notes.
14. Lifton-Zoline, notes.
15. Rabinowitz, unpublished interview with Youngblood.
16. Youngblood, unpublished draft.
17. Mobile Image, *Hole-in-Space* proposal.
18. Lifton-Zoline, notes.
19. Galloway, unpublished interview with Youngblood.
20. Rabinowitz, unpublished interview with Youngblood.
21. Lifton-Zoline, notes.
22. Mobile Image, unpublished interview with Youngblood.
23. Lifton-Zoline, notes.
24. Rabinowitz, unpublished interview with Youngblood.
25. Frederick Williams, *The Communications Revolution* (Beverly Hills, CA: Sage, 1982). Habermas's influential book *The Structural Transformation of the Public Sphere* was not published in the United States until 1989 (by the MIT Press), although excerpts and shorter versions of his argument as well as some of those of his critics appeared as early as 1974 (see, e.g., Jürgen Habermas, "The Public Sphere: An Encyclopedia Article," *New German Critique* 3 [Autumn 1974]: 49–55; Negt, "Mass Media," 61–80). The other voices alluded to here, and discussed in more detail later in the text, include Richard Sennett, *The Fall of Public Man* (New York: Alfred A. Knopf, 1977); Fredric Jameson, "Postmodernism, or, The Cultural Logic of Late Capitalism," *New Left Review* 146 (July–August 1984): 59–92; David Harvey, *The Condition of Postmodernity* (Malden, MA: Blackwell, 1989).
26. Paul Virilio, "The Overexposed City" (1984), reprinted in *The Blackwell City Reader*, ed. Gary Bridge and Sophie Watson (Hoboken, NJ: Wiley-Blackwell, 2002), 441, 444.
27. Hansen, "Foreword," xxx.
28. Castells is a prolific researcher and writer of critical urban studies, with numerous books and articles written during the late 1970s and 1980s alone. We are looking specifically to "Crisis, Planning, and the Quality of Life: Managing the New Historical Relationships between Space and Society," *Society and Space* 1 (1983): 3–21 and "High Technology, Economic Restructuring, and the Urban-Regional Process in the United States," in *High Technology, Space, and Society*, ed. Manuel Castells (Beverly Hills, CA: Sage, 1985), 11–40.
29. Castells, *High Technology*, 32.

30. Castells, *High Technology*, 17.

31. Castells, *High Technology*, 16–17.

32. Castells, *High Technology*, 18.

33. Castells, *High Technology*, 24.

34. Castells, "Crisis, Planning, and the Quality of Life," 7.

35. See Jürgen Habermas, *Legitimation Crisis* (Boston: Beacon Press, 1973); cited in Castells, "Crisis, Planning, and the Quality of Life," 10.

36. Castells, "Crisis, Planning, and the Quality of Life," 16.

37. Some of the key texts concerning this discussion include Rosalyn Deutsche's "Alienation in Berlin: Kirchner's Street Scenes," *Art in America* 71, no. 1(January 1983): 64–72; "Uneven Development: Public Art in New York City," *October* 47 (Winter, 1988): 3–52; and "Boys Town," *Society and Space* 9 (1991): 5–30; Craig Owens, "The Yen for Art," in *Discussions in Contemporary Culture*, ed. Hal Foster, Dia Art Foundation Discussions in Contemporary Culture, no. 1 (Seattle, WA: Bay Press, 1987), 16–23; Martha Rosler's project *If You Lived Here: The City in Art, Theory, and Social Activism*, a series of planning sessions, exhibitions, and discussions, convened in 1987–1989 and published under the same title by the Dia Art Foundation as part of its Discussions in Contemporary Culture series (Seattle, WA: Bay Press, 1991); and Miwon Kwon, "The Wrong Place," *Art Journal* 59, no. 1 (Spring 2000): 33–43.

38. Rosalyn Deutsche, "Agoraphobia," in *Evictions: Art and Spatial Politics* (Cambridge, MA: MIT Press, 1996), 283.

39. Edward Soja, *Postmodern Geographies: The Reassertion of Space in Critical Social Theory* (London: Verso, 1989); Michael Sorkin, ed., *Variations on a Theme Park: The New American City and the End of Public Space* (New York: Noonday Press, 1992).

40. Rosalyn Deutsche, "Boys Town" (1991), reprinted in *Evictions: Art and Spatial Politics* (Cambridge, MA: MIT Press, 1996), 210.

41. M. Christin Boyer, "The City of Illusion: New York's Public Places," in *The Restless Urban Landscape* (Englewood, NJ: Prentice-Hall, 1993), 113–114.

42. Tom Finkelpearl, "Introduction: The city as Site," in *Dialogues in Public Art* (Cambridge and London: MIT Press, 2000), 21.

43. For examples, see: *Creative Time: The Book* (New York: Princeton Architectural Press, 2007).

44. Richard Goldstein, "Enter the Anti Space," *The Village Voice*, November 5–11, 1980, reprinted in *Show and Tell: A Chronicle of Group Material*, ed. Julie Ault (London: Four Corners Books, 2010), 20.

45. Lucy Lippard, *The Lure of the Local: Senses of Place in a Multicentered Society* (1997), cited and critiqued by Kwon, "The Wrong Place," 34–35.

46. Kwon, "The Wrong Place," 39.

47. Jan Avgikos, "Group Material Timeline: Activism as a Work of Art," in *But Is It Art? The Spirit of Art as Activism*, ed. Nina Felshin (Seattle, WA: Bay Press, 1995), 99.

48. Kwon, "The Wrong Place," 34–36.

49. Sharon Zukin, *The Culture of Cities* (Cambridge, MA: Blackwell, 1995), 7–10.

50. Zukin, *The Culture of Cities*, 121.

51. M. Christine Boyer, *The City of Collective Memory: Its Historical Imagery and Architectural Entertainments* (Cambridge, MA: MIT Press, 1994), 46–47, 372–373.

52. Reyner Banham, *Los Angeles: The Architecture of Four Ecologies* (Baltimore, MD: Pelican Books, 1973), 151–156.

53. C. Ondine Chavoya and Rita Gonzalez, "Asco and the Politics of Revulsion," in *Asco: Elite of the Obscure, A Retrospective, 1972–1987* (Ostifildern, Germany: Hatje Cantz Verlag, 2011), 53–54.

54. Chavoya and Gonzalez, "Asco and the Politics of Revulsion," 46.

55. Amelia Jones, "Survey," in Tracey Warr, *The Artist's Body* (London: Phaidon, 2000), 34.

56. Ann Messner, statement in *A Book About Colab (and Related Activities)*, ed. Max Schumann (New York: Printed Matter, 2015), 104.

57. Jeff Kelley, "The Body Politics of Suzanne Lacy," in *But Is It Art? The Spirit of Art as Activism*, ed. Nina Felshin (Seattle, WA: Bay Press, 1995), 242.

58. Patricia C. Phillips, "Temporality in Public Art," in *Critical Issues in Public Art*, eds. Harriette Senie and Sally Webster (New York: Harper Collins, 1992), 303.

59. Kwon, "The Wrong Place," 34.

60. Writings of significant influence on US discourses of art and urbanism include Henri Lefebvre, *The Production of Space* (Malden, MA: Blackwell, 1991); Jane Jacobs, *The Death and Life of Great American Cities* (New York: Random House, 1961); Oskar Negt and Alexander Kluge, "*The Public Sphere and Experience*: Selections," *October* 46 (Fall 1988): 60–82.

61. Pollock, "Screening the Seventies," 155–199. For Brecht's reception in and influence on cinema and its studies as chronicled by *Screen*, see, for example, *Brecht and a Revolutionary Cinema*, *Screen* 15, no. 2 (Summer 1974) and *Brecht and the Cinema/Film and Politics*, *Screen* 16, no. 4 (Winter 1975/6).

62. Harvey, "Whose Brecht?," 45–59.

63. Michael W. Jennings and Tobias Wilke, "Editors' Introduction. Walter Benjamin's Media Tactics: Optics, Perception, and the Work of Art," *Grey Room* 39 (Spring 2010): 7. The publications alluded to include Walter Benjamin, "The Work of Art in the Age of Mechanical Reproduction" (1936–1939), in *Illuminations*, trans. Harry Zohn and ed. Hannah Arendt (New York: Schocken Books, 1969), 217–251; Susan Buck-Morrs, "Benjamin's *Passagen-Werk*: Redeeming Mass Culture for the Revolution," *New German Critique* 29 (Spring/Summer 1983): 211–240; Miriam Hansen, "Benjamin, Cinema, and Experience," *New German Critique* 40 (Winter 1987): 179–224; Benjamin's "allegorical method of construction" features prominently in Craig Owens' influential essay "The Allegorical Impulse: Toward a Theory of Postmodernism," *October* 12 (Spring 1980): 67–86.

64. Hansen, "Benjamin, Cinema, and Experience," 182.
65. Cited in Buck-Morrs, "Benjamin's *Passagen-Werk*," 214.
66. Buck-Morrs, "Benjamin's *Passagen-Werk*," 216.
67. Buck-Morrs, "Benjamin's *Passagen-Werk*," 223.
68. Buck-Morrs, "Benjamin's *Passagen-Werk*," 213.
69. Karl Marx, "Capital. Volume 1," in *The Marx-Engels Reader*, ed. Robert Tucker (New York: Norton, 1978), 321.
70. Deutsche, "Alienation in Berlin," 64–72. See also Deutsche's "Representing Berlin," in *Evictions: Art and Spatial Politics* (Cambridge, MA: MIT Press, 1996), 109–158.
71. Deutsche, "Alienation in Berlin," 69.
72. Deutsche, "Alienation in Berlin,"
73. Deutsche, "Alienation in Berlin," 71.
74. Deutsche, "Representing Berlin," in *Evictions*, 150.
75. Deutsche, "Representing Berlin," 150.
76. Castells, *High Technology*, 17.
77. Castells, *High Technology*, 17.
78. Mike Featherstone, "The *Flaneur*, the City and Virtual Public Life," *Urban Studies* 35, no. 5–6 (May 1998).
79. Jürgen Habermans, "The Public Sphere: An Encyclopedia Article (1964)," *New German Critique* 3 (Autumn 1974): 53.
80. Martha Gever, "Pressure Points: Video in the Public Sphere," *Art Journal* 45, no. 3 (Fall 1985): 238–243.
81. Tobias Wilke, "Tacti(ca)lity Reclaimed: Benjamin's Medium, the Avant-Garde, the Politics of the Senses," *Grey Room* 39 (Spring 2010): 40.
82. Wilke, "Tacti(ca)lity Reclaimed," 40 (authors' emphasis).
83. Hansen, "Benjamin, Cinema, and Experience," 182.
84. Krzysztof Wodiczko, "Public Projections," *October* 38 (Fall 1986): 10.
85. Deutsche, "Krzysztof Wodiczko's *Homeless Projection*," in *Evictions: Art and Spatial Politics* (Cambridge, MA: MIT Press, 1996), 78–89.
86. Grace Glueck, "And Now, a Few Words from Jenny Holzer" *New York Times*, December 3, 1989, 42.
87. Craig Owens, "The Discourse of Others: Feminists and Postmodernism" (1983), in *Beyond Recognition: Representation, Power, and Culture* (Berkeley: University of California Press, 1994), 181.
88. Owens, "The Discourse of Others," 184.
89. As Benjamin put it, "the bourgeois individual as flaneur could take delight in the 'crowd' precisely because it was not congealed into a revolutionary class" (Buck-Morrs, "Benjamin's *Passagen-Werk*," 230).

90. Featherstone, "The *Flaneur*," 919–920.

91. Walter Benjamin, *The Arcades Project* (Cambridge, MA: Belknap Press of Harvard University Press, 1999), 345.

92. Benjamin Buchloh, "From Gadget Video to Agit Video: Some Notes on Four Recent Video Works," *Art Journal* 45, no. 3 (Fall 1985): 224.

93. Buchloh, "From Gadget Video to Agit Video," 224.

94. Erkki Huhtamo, *Illusions in Motion: Media Archelology of the Moving Panorama and Related Spectacles* (Cambridge, MA: MIT Press, 2013). See also Huhtamo, "Screen Tests: Why Do We need an Archeology of the Screen?" *Cinema Journal* 51, no. 2 (Winter 2012): 144–148 and "Messages on the Wall: An Archeology of Public Media Displays," *Urban Screens Reader*, ed. Scott McQuire, Meredith Martin, and Sabine Niederer (Amsterdam: Institute of Network Cultures, 2009), 15–28.

95. Huhtamo, "Messages on the Wall," 15; Scott McQuire, "Mobility, Cosmopolitanism, and Public Space in the Media City," in *Urban Screens Reader*, ed. Scott McQuire, Meredith Martin, and Sabine Niederer (Amsterdam: Institute of Network Cultures, 2009), 47–48.

96. Andreas Broeckmann, "Intimate Publics: Memory, Performance, and Spectacle in Urban Environments," in *Urban Screens Reader*, ed. Scott McQuire, Meredith Martin, and Sabine Niederer (Amsterdam: Institute of Network Cultures, 2009), 109.

97. Saskia Sassen, "Reading the City in a Global Digital Age," in *Urban Screens Reader*, ed. Scott McQuire, Meredith Martin, and Sabine Niederer (Amsterdam: Institute of Network Cultures, 2009), 32.

98. Arjun Appadurai, *Modernity at Large: Cultural Dimensions of Globalizations* (Minneapolis: University of Minnesota Press, 1996), 23.

99. Benjamin Bratton, *The Stack: On Software and Sovereignty* (Cambridge, MA: MIT Press, 2015), 41–72. See also Bratton, "The Black Stack," *e-flux Journal* 53 (March 2014) and "Blur: On The Stack and The Stack-to-Come," *Fall Semester 2015*, https://www.librarystack.org/blur-on-the-stack-and-the-stack-to-come/.

100. Bratton, "The New Normal," *Strelka Mag*, accessed October 23, 2022, https://strelkamag.com/en/article/the-new-normal-essay-bratton.

101. Bratton, "The Black Stack."

102. Bratton, "The Black Stack."

103. Anna Everett, *Digital Diaspora: A Race for Cyberspace* (Albany: SUNY Press, 2009), 32.

104. Everett, *Digital Diaspora*, 31. "IT" here refers to both "information technologies" and political contestation/struggle; Everett recounts a 1997 action by the Black Geeks Online organization taking their "online literacy activism off-line to the streets of Washington, D.C. . . . host[ing] a one-day information technology expo to bring the 'Net to the un-connected.'"

105. Huhtamo, "Messages on the Wall," 20.

106. Huhtamo, "Messages on the Wall," 24–25.

107. Huhtamo, *Illusions in Motion*, 29–62.

108. McQuire, "Mobility, Cosmopolitanism, and Public Space in the Media City," 48.
109. McQuire, "Mobility, Cosmopolitanism, and Public Space in the Media City," 56.
110. *Body Movies* was first staged in Rotterdam (2001), followed by several other cities including Hong Kong (2006); *Underscan* was staged in numerous cities in the UK (2005–2006) and in Trafalgar Square, London (2008). See McQuire, "Mobility, Cosmopolitanism, and Public Space in the Media City," 57, fn. 33.
111. McQuire, "Mobility, Cosmopolitanism, and Public Space in the Media City," 58–59.
112. Wendy Hui Kyong Chun, "Net-munity, or the Space between Us . . . Will Open the Future," *Critical Inquiry* 47 (Winter 2021): S108.
113. Broeckmann, "Intimate Publics," 111.
114. Everett, *Digital Diaspora*, 18.
115. Everett, *Digital Diaspora*, 18–19.
116. Crary, "Géricault," 7–8.
117. McQuire, "Mobility, Cosmopolitanism, and Public Space in the Media City," 58.
118. McQuire, "Mobility, Cosmopolitanism, and Public Space in the Media City," 33.
119. Bratton, *The Stack*, 154.
120. Bratton, *The Stack*, 252.
121. Bratton, *The Stack*, 252
122. Deutsche, "Agoraphobia," 296.
123. Deutsche, "Agoraphobia," 294, 296.
124. Deutsche, "Agoraphobia," 298.
125. Deutsche, "Agoraphobia," 300.
126. Bratton, *The Stack*, 220.
127. Bratton, *The Stack*, 255.
128. Bratton, *The Stack*, 159.
129. Benjamin Bratton, *The Revenge of the Real: Politics for a Post-Pandemic World* (London: Verso, 2021), 72.
130. Bratton, *Revenge of the Real*, 74.
131. Kwon, "The Wrong Place," 42.
132. Alexander Kluge, *Gärten der Kooperation/Gardens of Kooperation* (Stuttgart, Germany: Württembergischer Kunstverein, 2017), 84 (authors' emphasis).

CHAPTER 3

1. Negt and Kluge, "*The Public Sphere and Experience:* Selections," 60.
2. David Harvey calls neoliberalism a "failed if not disingenuous and utopian project masking the restoration of class power," in "Neoliberalism as Creative

Destruction," *Annals of the American Academy of Political and Social Science* 610, (March 2007): 22–44.

3. Harvey, "Neoliberalism as Creative Destruction," 66.
4. Harvey, "Neoliberalism as Creative Destruction," 65.
5. Harvey, "Neoliberalism as Creative Destruction," 34–35.
6. Haraway, "Situated Knowledges"; bell hooks, "marginality as site of resistance," in *Out There: Marginalization and Contemporary Culture*, ed. Russell Ferguson et al. (Cambridge, MA: MIT Press, 1990), 341–343; Nancy Fraser, "Rethinking the Public Sphere: A Contribution to the Critique of Actually Existing Democracy," in *Habermas and the Public Sphere*, ed. Craig Calhoun (Cambridge, MA: MIT Press, 1992), 109–142; Michael Warner, "The Mass Public and the Mass Subject," in *Habermas and the Public Sphere*, ed. Craig Calhoun (Cambridge, MA: MIT Press, 1992), 377–401; Negt and Kluge, "*The Public Sphere and Experience:* Selections"; Eberhard Knodler-Bunte et al., "The Proletarian Public Sphere and Political Organization: An Analysis of Oskar Negt and Alexander Kluge's *The Public Sphere and Experience*," *New German Critique* 4 (Winter 1975): 51–75; Oskar Negt and Alexander Kluge, "Selections from Public Opinion and Practical Knowledge: Toward and Organizational Analysis of Proletariat and Middle Class Public Opinion," *Social Text* 25/26 (1990): 24–32.
7. Raúl H. Villa, "The Right to the City in Los Angeles: Discourse and Practice of a Chicano Alternative Public Sphere," in *Masses, Classes and The Public Sphere*, ed. Mike Hill and Warren Montag (London: Verso, 2000), 41–61; Deutsche, "Uneven Development"; Soja, *Postmodern Geographies*; Castells, "Crisis, Planning, and the Quality of Life"; Saskia Sassen, "The City: Localizations of the Global," *Perspecta* 36 (2005): 73–77.
8. Harvey, "Whose Brecht?," 45–59.
9. Mobile Image, "EC Original Development Materials," c.1983 (Rabinowitz/Galloway Archives).
10. Mobile Image, "EC Original Development Materials."
11. Negt and Kluge, cited in Jameson, "On Negt and Kluge," 156, fn. 5.
12. Negt and Kluge, "*The Public Sphere and Experience:* Selections," 62.
13. Mobile Image, "EC Original Development Materials."
14. Negt and Kluge, "*The Public Sphere and Experience:* Selections," 63.
15. Kiaer, "Boris Arvatov's Socialist Objects," 108.
16. Kiaer, "Boris Arvatov's Socialist Objects," 109, 116.
17. From *Electronic Café* proposal, c.1983, in "Miscellaneous" file (Rabinowitz/Galloway Archive).
18. Haraway, "Situated Knowledges," 579.
19. Haraway, "Situated Knowledges," 580.
20. Haraway, "Situated Knowledges," 584.
21. Haraway, "Situated Knowledges," 583.

22. Los Angeles Olympic Organizing Committee, Richard B. Perelman, editor-in-chief, *Official Report of the Games of the XXIII Olympiad, Los Angeles, 1984, Volume 1: Organization and Planning* (1985), 528.

23. Robert Fitzpatrick, "The Olympic Arts Festival," *Olympic Review* 198 (April 1984): 247.

24. Promotional materials for the Olympic Arts Festival (Rabinowitz/Galloway Archive).

25. Samuel Mark, letter of support for *Electronic Café* proposal (Rabinowitz/Galloway Archive).

26. Jose Luis Sedano, interview with the authors, Los Angeles, March 3, 2016.

27. Hye-Sook Park, interview with the authors, Los Angeles, March 2, 2016.

28. Gunther Hiller, quoted in Kenneth Fanucchi, "Café Owner Shuts Doors, Stays Hopeful," *Los Angeles Times*, April 11, 1985.

29. Sherrie Rabinowitz, quoted in Don Snowden, "Mobile Image Works on a Global Vision," *Los Angeles Times*, October 19, 1987.

30. Community Memory Project, informational materials, c.1982 (Rabinowitz/Galloway Archive).

31. *The Community Memory Project: An Introduction*, 1982, 12 (Rabinowitz/Galloway Archive).

32. *The Community Memory Project*. Galloway rewrote by hand the passages marked here in bold.

33. Community Memory unveiled their new network in Berkeley later that same summer. Terminals were located at the La Pena community center, the Whole Earth Access store, and the Telegraph Avenue Co-Op grocery store. The first went online on July 17, 1984, and the other two followed in August and September.

34. Manuel Castells, *The Rise of the Network Society* (Malden, MA: Blackwell, 1996), 5.

35. Fred Turner, *From Counterculture to Cyberculture* (Chicago: University of Chicago Press, 2006), 4.

36. Gene Youngblood, "Metadesign: Toward a Postmodernism of Reconstruction" *Ars Electronica* catalog (Linz, Austria: Linzer Veranstaltungsgesellschaft, 1986), n.p.

37. *Community Memory News*, newsletter, Number One, 1983 (Rabinowitz/Galloway Archive).

38. *The Community Memory Project*.

39. Galloway, interview with the authors, Piñon Hills, California, May 6, 2015.

40. Rabinowitz, unpublished interview (Rabinowitz/Galloway Archive).

41. Galloway, interview with the authors.

42. Galloway, interview with the authors.

43. Bob Smith, "Evolution of the Institute," *LAICA Journal* 1 (June 1974): 26.

44. Jane McFadden, "Here, Here, or There: On the Whereabouts of Art in the Seventies, in *Pacific Standard Time: Los Angeles Art 1945–1980*, exhibition catalog, ed.

Rebecca Peabody, Andrew Perchuck, Glenn Phillips, and Rami Singh (Los Angeles: Getty Research Institute and the J. Paul Getty Museum, 2011), 296.

45. Richard Armstrong, "Cultural Surplus," *LAICA Journal* 10 (March–April 1976): 17.

46. See, for example, "Observations," *LAICA Journal* 1 (June 1974): 22–23.

47. Smith, "Evolution of the Institute," 26.

48. Peter Frank, "Patterns in the Support Structure for California Art," *LAICA Journal* 19 (June–July 1978): 42.

49. Marcia Tucker, "Interview with Pat Steir," *LAICA Journal* 10 (March–April 1976): 23–24.

50. McFadden, "Here, Here, or There," 249–251.

51. Daniel Widener, *Black Arts West: Culture and Struggle in Postwar Los Angeles* (Durham, NC: Duke University Press, 2010), 8.

52. Widener, *Black Arts West*, 158–159.

53. Linda Frye Burnham, "The Late '70s Alternative: Artist-Run Art Spaces," in *Collaboration Labs: Southern California Artists and the Artist Space Movement*, exhibition catalog (Santa Monica, CA: 18th Street Art Center, 2011), 89–93.

54. McFadden, "Here, Here, or There," 258.

55. McFadden, "Here, Here, or There," 281.

56. Frank, "Patterns in the Support Structure," 43.

57. Frank, "Patterns in the Support Structure," 43.

58. McFadden, "Here, Here, or There," 259.

59. McFadden, "Here, Here, or There," 266.

60. Bryan-Wilson, "Out to See Video," 74.

61. Widener, *Black Arts West*, 12–13.

62. Burnham, "The Late '70s," 93.

63. Frank, "Patterns in the Support Structure," 42.

64. Dori Cypis, "Los Angeles: A Zone Beyond Time, A Personal Perspective on 1972–85," in *Collaboration Labs: Southern California Artists and the Artist Space Movement*, exhibition catalog (Santa Monica, CA: 18th Street Art Center, 2011), 97.

65. Anonymous user, post titled "The Gilligan's Island Syndrome," *Electronic Café* Community Memory record (Rabinowitz/Galloway Archive).

66. Anonymous user, post titled "THOUGHTS ABOUT ELECTRONIC CAFE," *Electronic Café* Community Memory record (Rabinowitz/Galloway Archive).

67. Anonymous user post, *Electronic Café* Community Memory record (Rabinowitz/Galloway Archive).

68. For an in-depth discussion of the "the popular" as a site of the political contest over meaning and identity, avoiding the ideological reproduction of the culturally "authentic," see Hall, "Notes on Deconstructing 'The Popular,'" 227–240.

69. Abraham F. Lowenthal, *Global California: Rising to the Cosmopolitan Challenge* (Stanford, CA: Stanford University Press, 2009), 54.

70. Roger Keil, *Los Angeles: Globalization, Urbanization and Social Struggles* (Chichester, UK: John Wiley, 1998), 9, 12, 13.

71. Widener, *Black Arts West*, 253.

72. Keil, *Los Angeles*, 21–27.

73. Keil, *Los Angeles*, 9–14.

74. Keil, *Los Angeles*, 120.

75. Ivan Light and Edna Bonacich, *Immigrant Entrepreneurs: Koreans in Los Angeles, 1965–1982* (Berkeley: University of California Press, 1988), 3–6.

76. Light and Bonacich, *Immigrant Entrepreneurs*; Donald Teruo Hata, Jr., and Nadine Ighitani Hata, "Asian-Pacific Angelinos: Model Minorities and Indispensable Scapegoats," in *20th Century Los Angeles: Power, Promotion, and Social Conflict*, ed. Norman M. Klein and Martin J. Schiesl (Claremont, CA: Regina Books, 1990), 85–89.

77. Light and Bonacich, *Immigrant Entrepreneurs*, 434–435.

78. Villa, "The Right to the City," 41–42.

79. Villa, "The Right to the City," 43–47.

80. Villa, "The Right to the City," 49–52.

81. Villa, "The Right to the City," 54.

82. Ricardo Romo, *East Los Angeles: History of a Barrio* (Austin: University of Texas Press, 1983), 171.

83. Rodolfo Acuña, *A Community Under Siege: A Chronicle of Chicanos East of the Los Angeles River, 1945–1975* (Los Angeles: Chicano Studies Research Center Publications, University of California, 1984), x–xi.

84. Acuña, *Community Under Siege*, 270.

85. Mike Davis, *City of Quartz: Excavating the Future in Los Angeles* (London: Verso, 1990/2006), 180–186.

86. Davis, *City of Quartz*, 270.

87. Davis, *City of Quartz*, 284.

88. Widener, *Black Arts West*, 241.

89. Davis, *City of Quartz*, 309.

90. Widener, *Black Arts West*, 254.

91. Widener, *Black Arts West*, 3, 234.

92. Widener, *Black Arts West*, 223–224.

93. Widener, *Black Arts West*, 247–248.

94. Davis, *City of Quartz*, 78.

95. Widener, *Black Arts West*, 12.

96. Widener, *Black Arts West*, 238.

97. Davis, *City of Quartz*, 78.

98. Mike Boehm, "MOCA trustee Peter Brant using his art to get business loans," *Los Angeles Times*, September 28, 2012.

99. Widener, *Black Arts West*, 239–240.

100. Jo-Anne Berelowitz, "A New Jerusalem: Utopias, MOCA, and the Redevelopment of Downtown Los Angeles," *Strategies* 3, 1990: 213.

101. Widener, *Black Arts West*, 234–235.

102. Anastasia Loukaitou-Sideris and Gail Sansbury, "Lost Streets of Bunker Hill," *California History* 74, no. 4 (Winter 1995/1996): 400–401.

103. Loukaitou-Sideris and Sansbury, "Lost Streets," 405.

104. Loukaitou-Sideris and Sansbury, "Lost Streets," 406.

105. "New Technology Permeates an Age-Old Tradition," *InfoWorld* 6, no. 31 (July 30, 1984): 27.

106. Perelman, *Official Report of the Games*, 43–56.

107. *InfoWorld*, "New Technology Permeates an Age-Old Tradition," 27 (Rabinowitz/Galloway Archive).

108. Ted Friedman, *Electric Dreams: Computers in American Culture* (New York and London: New York University Press, 2005), 99. Friedman characterizes the personalization of computing as quintessentially American, in contrast to the French Minitel system, which treated computing as a public utility and thus reached a much larger of the French population in the 1980s.

109. For a discussion of the reciprocal relationship between utopian and dystopian narratives in the twentieth century, see Tom Moylan, *Demand the Impossible: Science Fiction and the Utopian Imagination* (Bern: Peter Lang, 2014), 7–9.

110. Friedman, *Electric Dreams*, 111.

111. Friedman, *Electric Dreams*, 105–106.

112. Friedman, *Electric Dreams*, 106.

113. Friedman, *Electric Dreams*, 106.

114. Friedman, *Electric Dreams*, 100–101.

115. Turner, *From Counterculture to Cyberculture*, 2, 32.

116. Turner, *From Counterculture to Cyberculture*, 208–209.

117. Marc Ries, "Rendezvous: The Discovery of Pure Sociality in Early Net Art," in Daniels and Reisinger, *Net Pioneers 1.0*, 72–73.

118. Nora and Minc (1980), quoted in Roy Ascott, *Telematic Embrace: Visionary Theories of Art, Technology, and Consciousness*, ed. Edward A. Shanken (Berkeley, CA : University of California Press, 2003), 50–51.

119. Edward A. Shanken, *Telematic Embrace*, 52.

120. Roy Ascott, "Art and Telematics: Towards a Network Consciousness" (1984), in *Telematic Embrace: Visionary Theories of Art, Technology, and Consciousness*, ed. Edward A. Shanken (Berkeley, CA: University of California Press, 2003), 187.

121. Shanken, *From Cybernetics to Telematics*, 51.

122. Robert Adrian X, quoted in Reinhard Braun, "From Representation to Networks: Interplays of Visualities, Apparatuses, Discourses, Territories, and Bodies," trans. Cecilia White, in *At a Distance*, 79.

123. Adrian X, "The World in 24 Hours," Ars Electronica catalog entry, accessed May 24, 2019, http://90.146.8.18/en/archives/festival_archive/festival_catalogs/festival_artikel.asp?iProjectID=12945.

124. Adrian X, "The World in 24 Hours."

125. Adrian X and Eric Gidney, *The World in 24 Hours* project documentation, included as addendum to Gidney's MA thesis "Artists' Use of Interactive Telephone-based Communications Systems from 1977–1984" (City Art Institute, Sydney, 1986), 120.

126. Haraway, "Situated Knowledges," 588.

127. Haraway, "Situated Knowledges," 584. This assessment of *Signal Breakdown* and its connection to Haraway derives from an unpublished paper by Erin Dickey, "'Where are the Women?': Positioning *Signal Breakdown* in *The World in 24 Hours*," May 5, 2019, pp. 3–4.

128. Fredric Jameson, "Progress versus Utopia; Or, Can We Imagine the Future?" *Science Fiction Studies* 9, no. 2 (July 1982): 147–158; Donna Haraway, "A Manifesto for Cyborgs: Science, Technology, and Socialist Feminism" and "A Cyborg Manifesto: Science, Technology, and Socialist-Feminism in the Late 20th Century" (1983/4), in *Simians, Cyborgs, and Women: The Reinvention of Nature* (London: Routledge, 1991), 141–181; Tom Moylan, *Demand the Impossible: Science Fiction and the Utopian Imagination*, ed. Raffaella Baccolini and Tom Moylan (Bern: Peter Lang, 2014) (original work published by Tom Moylan in 1986); *Black American Literature Forum* 18, no. 2, Science Fiction Issue (Summer,1984).

129. Moylan, *Demand the Impossible*, 1.

130. Moylan, *Demand the Impossible*, 3.

131. This history is described most succinctly in Moylan, *Demand the Impossible*, 1–8.

132. Moylan, *Demand the Impossible*, 4.

133. Robert Elliot Fox, "The Politics of Desire in Delaney's *Triton* and *The Tides of Lust*," *Black American Literature Forum* 18, no. 2 (Summer 1984): 52.

134. Moylan, *Demand the Impossible*, 10.

135. Moylan, *Demand the Impossible*, 10.

136. Cited in Mary Kay Bray, "Rituals of Reversal: Double Consciousness in Delaney's *Dhalgren*," *Black American Literature Forum* 18, no. 2 (Summer 1984): 57.

137. Galloway and Rabinowitz, accessed February 24, 2015, http://www.ecafe.com/getty/ART/index.html.

138. Galloway and Rabinowitz, lecture video, 1988.

139. Mobile Image, "EC Original Development Materials," c.1983.

140. Negt and Kluge, "*The Public Sphere and Experience:* Selections," 65.

141. Negt and Kluge, "*The Public Sphere and Experience:* Selections," 63.

142. Félix Guattari, *Molecular Revolution: Psychiatry and Politics* (1977), trans. Rosemary Sheet (Middlesex, UK: Penguin Books, 1984), 236.

143. Guattari, *Molecular Revolution*, 241. Guattari continues: "It is not necessary at this stage to produce blueprints for a substitute society."

144. Galloway, interview with the authors.

145. For a historical account of the public sphere and its institutions, see Jürgen Habermas, *The Structural Transformation of the Public Sphere: An Inquiry into a Category of Bourgeois Society*, trans. Thomas Burger (Cambridge, MA: MIT Press, 1991).

146. Hansen, "Unstable Mixtures," 201.

147. Schiller, *Communication and Cultural Domination.*

148. Schiller, *Communication and Cultural Domination*, 49–50.

149. Schiller, *Communication and Cultural Domination*, 89.

150. Youngblood, "Metadesign," n.p.

151. Mary P. Ryan, "Gender and Public Access: Women's Politics in Nineteenth-Century America," in *Habermas and the Public Sphere*, ed. Craig Calhoun (Cambridge, MA: MIT Press, 1992), 259; Fraser, "Rethinking the Public Sphere"; Warner, "The Mass Public and the Mass Subject."

152. Ryan, "Gender and Public Access," 283–284.

153. Fraser, "Rethinking the Public Sphere," 132.

154. Fraser, "Rethinking the Public Sphere," 123.

155. Haraway, "Situated Knowledges," 579 (emphasis in original).

156. Haraway, "Situated Knowledges," 575.

157. Haraway, "Situated Knowledges," 581 (emphasis in original).

158. See, for example, hooks, "marginality as site of resistance"; Trinh T. Minh-Ha, *Woman, Native, Other: Writing Postcoloniality and Feminism* (Bloomington: Indiana University Press, 1989); Gloria Anzaldúa, *Borderlands/La Frontera: The New Mestiza* (San Francisco: Spinsters/Aunt Lute, 1987); Homi Bhabha, *The Location of Culture,* Op. Cit.

159. hooks, "marginality as site of resistance," 341.

160. hooks, "marginality as site of resistance," 341

161. hooks, "marginality as site of resistance," 342.

162. bell hooks, *Yearning: Race, Gender, and Cultural Politics* (Boston: South End Press, 1990), 149, 20.

163. Bhabha, *The Location of Culture*, 1–2.

164. Negt and Kluge, "*The Public Sphere and Experience:* Selections," 60.

165. Haraway, "Situated Knowledge," 584.

166. Sassen, "The City," 73 (authors' emphasis).

167. Sassen, "The City," 73–74.

168. Sassen, "The City," 76.

169. The term *Umfunktionierung* (refunctioning) is Bertolt Brecht's, referring to the strategic, revolutionary inversion of technological devices. See, for example, "The Radio as an Apparatus of Communication," 51–53.

170. Peter Stallybrass and Allon White, *The Politics and Poetics of Transgression* (Ithaca, NY: Cornell University Press, 1986).

171. Stallybrass and White, *The Politics and Poetics of Transgression*, 96.

172. Stallybrass and White, *The Politics and Poetics of Transgression*, 97, 99.

173. Sassen, "The City," 75.

174. Stallybrass and White, *The Politics and Poetics of Transgression*, 99.

175. Hunter Drohojowska, "Holograms document the faces of America," *Los Angeles Herald Examiner*, June 29, 1984, 31.

176. Susan Otto, "Cultural Production and Popular Mechanics: Projects 1978–95," in *Daniel J. Martinez: The Things You See When You Don't Have a Grenade*, ed. David Levi Strauss et al. (Santa Monica, CA: Smart Art Press, 1996), 91.

177. Martinez, interview with the authors.

178. Donna Haraway, "A Cyborg Manifesto: Science, Technology, and Socialist-Feminism in the Late 20th Century" (1983/4), in *Simians, Cyborgs, and Women: The Reinvention of Nature* (London: Routledge, 1991), 151.

179. Haraway, "Cyborg Manifesto," 150–151.

180. Otto, "Cultural Production and Popular Mechanics," 91.

181. Haraway, "Cyborg Manifesto," 153.

182. Haraway, "Cyborg Manifesto," 153.

183. Haraway, "Cyborg Manifesto," 152.

184. Roland Barthes, "Myth Today" (1957), in *Mythologies* (New York: Hill and Wang, 1972), 107–159. To Barthes, "myth is depoliticized speech" and hence, in this case, a way to articulate, present, and utilize culture and representation, the "local" and its constituencies as fixed and homogeneous rather than constructed, complex, and heterogeneous: "What the world supplies to myth is a historical reality, defined, even if it goes back quite a while, by the way in which men have produced or used it; and what myth gives in return is a *natural* image of this reality" (142).

185. Jenkins cited in Rebecca Peabody, "African American Avant-Gardes, 1965–1990," *Getty Research Journal* 1 (2009): 214.

186. Hammer Museum, "Ulysses Jenkins, *Dream City*," Digital *Archive Now Dig This!: Art and Black Los Angeles 1960–1980*, accessed November 15, 2019, https://hammer.ucla.edu/now-dig-this/art/dream-city.

187. Jenkins, phone interview with the authors, September 17, 2019.

188. Jenkins, phone interview with the authors.

189. Aria Dean, "Written and Bitten: Ulysses Jenkins and the Non-Ontology of Blackness," *X-TRA* 19, no. 2 (Winter 2017): 11.

190. Dean, "Written and Bitten," 8, 12.

191. Dean, "Written and Bitten," 16 (authors' emphasis). The phrase "consent not to be a single being" appears often in the writings of Fred Moten who, in turn, references Édouard Glissant. See Dean, "Written and Bitten," fn. 24.

192. Crary, "Géricault," 7–8.

193. Stallybrass and White, *Politics and Poetics of Transgression*, 93–96.

194. Jean Baudrillard, "Ecstasy of Communication," trans. John Johnston, in *The Anti-Aesthetic: Essays on Postmodern Culture*, ed. Hal Foster (Seattle, WA: Bay Press, 1983), 127–133.

195. Jean-François Lyotard, *The Postmodern Condition: A Report on Knowledge*, trans. Geoff Bennington and Brian Massumi (Minneapolis: University of Minnesota Press, 1984), 66–67.

196. Jameson, "Postmodernism," 115.

197. Jameson, "Postmodernism," 115.

198. See, for example: Thomas Lawson, "Last Exit: Painting," *Artforum* 20, no. 2 (October 1981): 40–47.

199. Jean-François Lyotard, interview in *Flash Art* 121 (March 1985), reprinted in: http://www.art-agenda.com/reviews/les-immateriaux-a-conversation-with-jean-francois-lyotard-and-bernard-blistene/.

200. Alexander Alberro, "Reconsidering Conceptual Art, 1966–1977," in *Conceptual Art: A Critical Anthology*, ed. Alexander Alberro and Blake Stimson (Cambridge, MA: MIT Press, 2000), xvii.

201. Lucy Lippard and John Chandler, "The Dematerialization of Art" (1968), reprinted in Alberro and Stimson, eds., *Conceptual Art*, 47–48.

202. Sol LeWitt, "Paragraphs on Conceptual Art" (1967), reprinted in Alberro and Stimson, eds., *Conceptual Art*, 12.

203. Benjamin Buchloh, "Conceptual Art 1962–1969: From the Aesthetic of Administration to the Critique of Institutions" (1989), reprinted in Alberro and Stimson, eds., *Conceptual Art*, 530–533.

204. Alberro, "Reconsidering Conceptual Art," xxix.

205. Alberro, "Reconsidering Conceptual Art," xxx.

206. Negt and Kluge, "*The Public Sphere and Experience:* Selections," 72.

207. Gene Youngblood, "Virtual Space: The Electronic Environments of Mobile Image," *International Synergy Journal* 1, no. 1 (1986), accessed January 20, 2023, http://www.ecafe.com/museum/is_journal/is_journal.html.

208. For an in-depth discussion of the relevance of Negt and Kluge's concept of counterpublicity for post-1960s US politics of social subjectivity and solidarity, see Hansen, "Unstable Mixtures," 207.

209. Daniels, "Inventing and Re-Inventing Radio," 41.

210. Galloway, interview with the authors.

211. Tetsuo Kogawa, "What Is Mini FM," accessed July 25, 2019, https://anarchy.translocal.jp/radio/micro/what_is_minifm/index.html.

212. Frantz Fanon, "This is the Voice of Algeria," in *A Dying Colonialism* (New York: Grove Press, 1965), 84.

213. Philip Glahn, "The Radio and/as Digital Productivism," in *Radio as Art: Concepts, Spaces, Practices*, ed. Anne Thurmann-Jajes et al. (Bielefeld, Germany: Transcript, 2019), 206.

214. Other stations included Radio Libera in Florence, Radio Emmanuel in Ancona, and Radio Milano International (Cosetta Gaudenzi, "Luciano Ligahue's Radiofreccia: Regionalism and Globalization." *Italica* 86, no. 2 (2009): 300).

215. Franco Berardi, interview with Rosetta Brooks, *Zg Press*, July 10, 2010, http://www.zgpress.com/?p=36.

216. Felix Guattari, "Millions and Millions of Potential Alices," in *Molecular Revolution: Psychiatry and Politics*, trans. Rosemary Sheed (New York: Penguin, 1984), 236–241.

217. Glahn, "Radio and/as Digital Productivism," 207.

218. Haraway, "Situated Knowledges," 587.

219. Shannon Kelly, "Introduction to the Program," in *LA Rebellion: Creating a New Black Cinema*, exhibition catalog (Los Angeles: Getty Foundation, 2011), 4.

220. Ben Caldwell, interview with the authors, Los Angeles, California, March 2, 2016.

221. Caldwell, quoted in Don Snowden, "Ben Caldwell's 'Fresh' Approach to Film Making," *Los Angeles Times*, January 8, 1988. Regarding Caldwell's early interest in Brecht, see Marcyliena Morgan, *The Real Hiphop: Battling for Knowledge, Power, and Respect in the LA Underground* (Durham, NC: Duke University Press, 2009), 39–42.

222. Sahra Sulaiman, "Fighting Gentrification with Sankofa Red—A Repurposed Pay Phone," Streetsblog LA, December 4, 2014, accessed June 26, 2020, https://la.streetsblog.org/2013/12/04/can-a-re-purposed-payphone-stave-off-gentrification-in-leimert-park/.

223. Rabinowitz, quoted in Anna Couey, "Restructuring Power: Telecommunications Works Produced by Women," in *Women, Art and Technology*, ed. Judy Malloy (Cambridge, MA: MIT Press, 2003), 61.

224. Mobile Image, "Electronic Cafe International-HQ Highlights," artists' website: http://www.ecafe.com/museum/history/ksoverview2.html.

225. Ana Coria, interview with the authors, Long Beach, California, September 16, 2015.

226. Wendy Smith, ". . . And the Kitchen Goes Electronic," *New York Times*, November 6, 1994.

227. Carl Loeffler, "The Art Com Electronic Network," *Leonardo* 21, no. 3 (1988): 320.

228. Wolfgang Staehle, 1994, quoted in Daniels, "Reverse Engineering," 18, n. 10.

229. Ries, "Rendezvous," 74; Daniels, "Reverse Engineering," 39.

230. Daniels, "Reverse Engineering," 32, n. 27.

231. Nicolas Bourriaud, *Postproduction: Culture as Screenplay: How Art Reprograms the World*, trans. Jeanine Herman (New York: Lukas & Sternberg, 2000), 8.

232. Nicholas Bourriaud, "Relational Aesthetics" (1998), trans. David Macey, reprinted in *Participation*, ed. Claire Bishop (London and Cambridge, MA: Whitechapel and MIT Press, 2006), 165.

233. Bourriaud, "Relational Aesthetics," 163.

234. Bourriaud, "Relational Aesthetics," 163.

235. Claire Bishop, "Antagonism and Relational Aesthetics," *October* 110 (Fall 2004): 58.

236. Radical Culture Research Collective (Gene Ray), "A Very Short Critique of Relational Aesthetics," accessed January 27, 2021, https://transform.eipcp.net/correspondence/1196340894/print.html.

INDEX

Note: Page numbers set in italics indicate a figure on that page.